AF539224

Essential Oil Plants and their Cultivation

Essential Oil Plants and their Cultivation

Dhiraj Acharya

RANDOM PUBLICATIONS
NEW DELHI - 110 002 (INDIA)

Essential Oil Plants and their Cultivation

ISBN 978-93-51112-80-8

Published in 2014 in India by

Reprint 2021

RANDOM PUBLICATIONS

4376-A/4B, Gali Murari Lal, Ansari Road

New Delhi-110 002

Phone: +9111-43580356, 23289044

E-mail: randomexports@gmail.com; sales@randompublications.com; info@randompublications.com

Type Setting by: Friends Media, Delhi-110089

Printed at : Mehra Printers, Delhi-110092

Preface

The essential oils industry is a very exciting field when handled correctly and it can be very profitable. There are however many pitfalls that can cost a farmer dearly, farmers who intend to grow essential oil plants must be aware of the following. The correct genetic plant material for the plant extract industry must be used and this requires appropriate soil, local climate conditions and water quality. Specialised planters and soil cultivation equipment are required as well as weeding and tilling equipment, harvesting equipment and extraction machinery. In addition to the equipment and suitable growing conditions, the grower must have high quality control, marketing intelligence i.e. quantity and quality, Capital, Enthusiasm, and enjoy hard work. Cultivation and processing of aromatic and medicinal plants including mushrooms have opened new opportunities for income generation in rural sector. Essential oils are highly concentrated, volatile oils that can be extracted from aromatic plants. Their use dates back to ancient times, and their wide variety of therapeutic, medicinal and culinary uses has ensured their continued popularity. About 700 different kinds of plants contain useful essential oils, and there are several methods employed to extract them, the most common of which is distillation. While essential oils can be very expensive to buy, they are relatively cheap to distill at home. This guide provides basic instructions on how to extract the oils using the relatively simple and effective water-and-steam distillation process. The quantity of essential oils contained in a plant varies over the course of the plant's development, so it is essential to harvest at the right time. This will depend on the type of plant, so you need to do some research to determine when to harvest. It is also critical to harvest the plants correctly - careless handling, harvesting the wrong parts, even harvesting at the wrong time of day can reduce the quantity and quality of the essential oils.

Essential oils are generally extracted by distillation, often by using steam. Other processes include expression or solvent extraction. They are used inperfumes, cosmetics, soaps and other products, for flavouring food and drink, and for adding scents to incense and household cleaning products. Essential oils have been used medicinally in history. Medical applications proposed by those who sell medicinal oils range from skin treatments to remedies for cancer and often are based solely on historical accounts of use of essential oils for these purposes. Claims for the efficacy of medical treatments, and treatment of cancers in particular, are now subject to regulation in most countries. As the use of essential oils has declined in evidence-based medicine, one must consult older textbooks for much information on their use. Modern works are less inclined to generalize; rather than refer to "essential oils" as a class at all, they prefer to discuss specific compounds, such as methyl salicylate, rather than "oil of wintergreen". Interest in essential oils has revived in recent decades with the popularity of aromatherapy, a branch of alternative medicine that claims that essential oils and other aromatic compounds have curative effects. Oils are volatilized or diluted in a carrier oil and used in massage, diffused in the air by anebulizer, heated over a candle flame, or burned as incense. The earliest recorded mention of the techniques and methods used to produce essential oils is believed to be that of Ibn al-Baitar (1188–1248), an Andalusian physician, pharmacist and chemist.

Hence, this well-knitted; well-researched and well-debated work on this subject is highly academic and research quality will be appreciated and accepted by students; scholars and teachers.

I thank all members of my team who have helped in the preparation of the book. My special thanks go to "Random Publications" who have published the book.

—Dhiraj Acharya

Contents

1

Organic Essential Oils

Organic essential oils are derived from plants that have been grown without the use of pesticides, on land that has been certified by an authorised regulatory agent such as ECOCERT or the Soil Association. Today, many aromatherapists and nurses prefer to use organic essential oils in their clinical treatments because they believe they have more healing power and vitality than conventional essential oils. There is also the issue of pesticide residues to be considered too, since they have far-reaching effects for both the environment, and our bodies.

Every one of our Organic Essential Oils has been analytically tested for purity and certified under one of the following official regulatory agencies:

- The Soil Association
- Nature et Progres
- ECOCERT
- Qualite-France SA
- Agrobio

Quinessence organic essential oils bring you unrivalled value for money because in most instances we have purchase them directly from the farms where they are produced, thereby cutting out the middle-man. We then pass these savings along to you, and in many cases our organic essential oils are not a great deal more expensive than many of our conventionally produced essential oils.

Further to the organic essential oils we purchase from around the world, an increasing selection of our Certified Organic essential oils are now distilled on-site from medicinal herbs and plants that have been grown for Quinessence on a 500 acre farm in the United Kingdom.

Whilst our preference is for organically produced essential oils, we do accept that it has yet to be proven scientifically they are any more effective than their conventionally produced counterparts. But even if it was proven there is no difference between them, we would still not change our view on this matter. There is much more to this subject than just efficacy. By purchasing organic essential oils from growers who use traditional farming practices, we are all contributing towards a more sustainable ecological environment for the future. Surely this is a sensible and worthwhile investment for our forthcoming generations? A large amount of agricultural land has already been lost due to soil erosion, and in many places the overuse of aggressive agrochemicals has destroyed the delicate balance between wildlife and its natural habitat. Discover more about the sources of Quinessence organic pure essential oils and the importance of buying organic by browsing other pages under this category.

Organic Status

To be certified as organic in the United Kingdom, a farm must first register with the Soil Association and then enter a 3 year conversion period. After this 3 year period an inspection is made by the Soil Association to establish that all regulations have been strictly observed. Only then is the farmer allowed to legally claim the product 'certified organic'.

The Quinessence range of Certified Organic essential oils are required to meet the following strict criteria:

- All crops have been botanically authenticated
- No chemical fertilisers - only farmyard manure used
- No chemical insecticides or fungicides were used
- Weeds were controlled by hand weeding

The term 'Organic' is defined by Law, and within the European Union organic farming it is governed by the European Council Regulation (EEC) No 2092/91 and means that farmers must abide by a strict set of rules. Organic status is only awarded to producers who have been inspected and shown to comply with all the necessary organic standards.

For almost 15 years we have been building close working relationships with growers, distillers and suppliers to ensure that our pure essential oils are produced by plants from a certified botanical species that were not subjected to the use of pesticides or herbicides. This philosophy is continuing to develop with an ever expanding number of producers who share our convictions with regards to organic agriculture and biodiversity. And this is precisely why we develop close working relationships with those growers who use traditional farming practices - to promote a more sustainable ecological environment for the future. Nowhere is this more true than in the United Kingdom, where we have forged allegiances with a network of farmers who are all dedicated to growing high quality, organic medicinal plants for the production of essential oils.

Sources and Origins of our Oils

Sourcing and purchasing organic and conventional essential oils from remote areas around the world is not a job for either a novice or the faint-hearted, because mistakes can be very, very costly for a company, in terms of both money and loss of reputation. Financial losses can be regained, but company reputations tarnished due to supplying poor quality oils are not so easily recovered. So we work hard all year round to ensure we preserve the quality and continuity of supply that our customers have come to expect. Keeping the supply chain as short as possible, cutting out the middlemen and meeting the people who grow the crops are all steps we must take to maintain our high standards.

It takes many years of training and practice to learn about the variability in therapeutic properties and the organic chemistry between a given species of plant. When we are procuring our oils, our search is targeted for those with high levels of bioactivity and a fine fragrance. Many essential oils have been 'adjusted' to meet the requirements of the perfumery or flavours industries, and whilst that may be acceptable for their requirements, it is certainly not for ours.

Growing Conditions

True experts at sourcing these natural raw materials know precisely where the finest herbs and essential oils are produced, but it takes many years to acquire this expertise and to also build up relationships with the producers and distilleries. In reality, few companies manufacturing or supplying aromatherapy products today have been in business long enough to have gained such vital expertise.

The fertility of the soil the plant was grown in, the genetic differences in the plant, the variety, cultivation practices, the climate, post-harvest handling and level of expertise used in the process of extraction will all have a significant effect on the resulting fragrance and bioactivity of an essential oil. This is why the country of origin is so important when it comes to the quality of an oil, since the climatic and soil conditions greatly affect the oil producing cells within the plant.

Lavender oil (Lavandula angustifolia) from France is a very good example; if you have several bottles of lavender from different suppliers try comparing them with one another, and note the differences. They will undoubtedly differ in aroma, chemical composition and therapeutic activity, and this can be for a number of reasons. These variations can be due to the altitude the plant was grown at, whether it was extracted from a population or clonal variety, or if it was produced organically. But there can be other reasons for variability as well.

Deception and Adulteration

It is quite common for lavender oil claiming its country of origin as France to have actually originated in Bulgaria or Croatia. Or it could be a mixture of oils from all three locations. And just to be clear; there is nothing wrong with the lavender oils produced in these countries - oils of excellent quality are produced in both regions, but they each have a different chemistry, odour profile and therapeutic action to that of French lavender. And a different bulk purchasing price that is considerably less than that of French lavender oil too!

Of course variations in quality can be due to blatant adulteration with synthetic or isolated chemicals, and this accounts for quite a large amount of poor quality essential oils that find their way into aromatherapy. This is usually the case with oils that have been adjusted to meet the required profile of the perfumery industry. But what is merely 'adjustment' or 'sophistication' for one industry can be adulteration to another. Interfering with the chemistry of an oil to meet an odour profile may suit the perfumer, but it does no favours to the aromatherapist. To obtain classic textbook results you must use pure essential oils as nature intended or they will not be as therapeutically effective.

There can be absolutely no compromises with quality, and at Quinessence we remain committed to only sourcing the very finest that nature has to offer, extracted from organically grown products from ecologically sustainable sources. Of course this is not at all an easy task, but during the past 20 years we have set our own standards of excellence from which we never vary. In today's fast-buck world, few other companies seem prepared to invest the time and money required to reach such exacting standards.

To discover more about the origins of Quinessence essential oils, you may like to visit the pages below.

- Chamomile Roman oil - United Kingdom
- Lavender oil - France
- Neroli oil - Tunisia
- Rose Otto - Bulgaria
- Yuzu oil - Japan

Tunisian Neroli Essential Oil

Neroli essential oil is extracted from the fragrant blossoms of the bitter orange tree (Citrus aurantium sub.sp amara), and has a beautiful aroma that appeals to men and women alike. In common with rose and jasmine, neroli oil is almost a complete fragrance in itself and forms the heart of one of the worlds most enduring perfumes, 'Eau de Cologne'.

Although neroli oil is produced in many countries such as Algeria, Egypt, France, Haiti, Italy, Morocco and Spain, the oils produced in France and Tunisia have always been considered to be the very finest and still command the highest price. At Quinessence, our preference is for Tunisian neroli oil.

Unlike much of its French counterpart, Tunisian bitter orange trees are not subjected to agrochemicals, simply because they are just too expensive for the smaller farmers to afford. Surprisingly, much of the neroli essential oil exported from Tunisia is produced from the blossoms of trees grown by small growing cooperatives and families, rather than from large-scale cultivation farms.

Origins and Folklore

It is believed that C. aurantium originated in South-East Asia, later spreading to North-Eastern India, Burma and China, and eventually finding its way via Arab traders to Africa, Arabia and Syria. From these regions it was taken to the Mediterranean by the Moors, and by the end of the 12th century it was cultivated in Seville, Spain, thereby leading to the common name for bitter oranges. It is not known precisely when or where the oil was first extracted by steam distillation, but legend has it that during the 17th century in Italy, Anne Marie de la Tremoille (Orsini), who was duchess of Bracciano and princess of Nerola, first introduced neroli oil as a fashionable fragrance to high society. She used it whilst bathing and also to perfume her stationary, scarves and most famously, her gloves.

The fragrance obviously caught on, because in 1709, the Italian perfumer J. M. Farina launched his blend of neroli, bergamot, lavender, lemon, petitgrain and rosemary onto an unsuspecting world, naming it 'Eau de Cologne'. The rest as they say, is history.

Fragrant Assets

The bitter orange tree is a small evergreen that typically reaches a height of 3 metres (10ft) in cultivation, but may attain up to 6 metres (20ft) when growing in the wild. It has a smooth brown trunk, stout branches and flexible green twigs with rather blunt thorns, and has a more erect stature and compact crown than that of the sweet orange tree (Citrus sinensis). The green twigs contain a significant amount of essential oil, and these along with the broad, ovate, glossy and highly aromatic leaves are used to produce petitgrain oil. The golden-yellow sour fruits are round or oval with a thick, heavily pitted skin that yield bitter orange essential oil by cold expression. This is a truly fragrant tree!

Critical Timing

Throughout April and May in Tunisia, prolific clusters of attractive, highly scented blossoms consisting of 5 petals and 24 yellow stamens begin to appear on the tree. Further to the beautiful oil within, these flowers also produce tasty nectar which proves irresistible to honey bees. The oil content of the flowers increases as they develop and bloom. The flowering buds are usually harvested by hand early in the morning just as they begin to open - but only on warm, sunny days, because damp or overcast weather can have an adverse effect on the fragrance of the oil. It is absolutely critical that the buds are collected at the correct stage of maturity, because if they are plucked too soon the yield of oil will be lower and the resulting oil will have an unwanted 'green' note in the fragrance.

Conversely, if the buds have opened too far when they are gathered much of the precious volatile oil will evaporate during the process of transportation to the stills. Getting the timing wrong either way will have a severe detrimental effect on both the fragrance and value of the distilled essential oil.

Preparation and Extraction

Before transportation to the distillery, the collected buds must be carefully 'winnowed' by hand to remove all traces of stray leaves, twigs or similar plant debris. Failure to remove this unwanted material will again result in a tainted fragrance of sub-standard quality, thereby be unable to command the high price normally associated with Tunisian neroli.

Isolation of the essential oil is achieved by low-pressure 'cool' water-steam distillation which yields a pale yellow oil with an exquisite, fresh, fruity-floral aroma. Up to 20% of the essential oil is dissolved into the water during the process of distillation, and this is recovered by solvent extraction resulting in what is known as 'orange flower water absolute'. Since this material consists mainly of the water-soluble components of the oil, the aroma is not very representative of the flower, but nontheless is still put to good use in the perfumery industry. Solvent extraction of the flowers produces a dark orange/brown viscous absolute with the most beautiful rich, warm and floral fragrance that closely resembles that of the blossoms.

Perfect for Skin

Appealing in aroma to both men and women, neroli is one of nature's most effective antidepressant oils, which together with its balancing and sedative properties makes it perfect for treating all types of negative emotional conditions, states of anxiety, menopause, and insomnia. This action may also in part explain the oils reputation as an effective aphrodisiac. Most problem skin conditions respond extremely well to treatments incorporating neroli, but for best results mix it with a hypoallergenic carrier lotion or base cream rather than a carrier oil. Used as part of a regular skin care routine, neroli improves elasticity, stimulates new cell growth, reduces thread veins, softens wrinkles and scars, and smells absolutely divine. The perfect skin care oil!

Bulgarian Rose Otto

The origin of the cultivated rose is often quoted as the Gulf of Persia, which is now known as Iran. From the 10th to the 17th century the rose industry was developed and dominated by Persia, and particularly in Shiraz, the famous city of poets and oriental culture.

From here the rose industry spread into Arabia, Mesopotamia, Palestine, Asia Minor (Anatolia) Greece, India, North Africa, and due to the conquering Moors reached as far as Spain. According to legend, in the 13th century R. damascena was brought from Damascus to Southern France by the returning Crusaders, although some experts believe it may have actually been R. gallica instead.

Valley of the Roses

During the 16th century, Ottoman (Turkish) merchants imported R. damascena for cultivation throughout the Balkan countries, including a newly founded town in Bulgaria that would eventually become known as Kazanlik. Kazan is the Turkish word for "still", and Kazanlik literally means 'the place of stills'. A nearby valley provided the perfect environment for growing roses, thereby establishing what would in later years become the finest rose oil producing region in the world. This area is now called the 'Valley of the Roses', and during 1878 cuttings from the improved stock were returned to Anatolia and planted in Isparta and Burdur, where current rose production still thrives.

Throughout the 19th century, the Bulgarian rose oil industry reigned supreme, almost monopolising the entire world supply of rose oil. This monopoly would not be broken until the industry was nationalised due to dramatic changes in the political and economical climate after World War 2, when production fell into a steady decline. Today, it is believed that Turkey holds the record as the largest producer of rose otto, and only the oil from this country matches the quality and fine fragrance to that of Bulgaria.

Climatic Conditions

In Bulgaria, the rose blossoms of Rosa damascena begin to bloom around the third week of May, and will continue for three or four weeks depending upon climatic conditions. The yield of oil can be dramatically affected by the prevailing weather conditions - for example during very hot and dry weather the harvest may last only two weeks and the yield of the oil is lowered due to loss by evaporation. Conversely, during mild and humid weather the harvest time can be extended whilst at the same time increasing the oil yield.

The harvesting season starts as soon as the flowers begin to open and continues until all the roses have been gathered. In Bulgaria and Turkey the blossoms are still collected by hand in the time-honoured way, and are nipped just below the calyx (the green, outer protective cover). Collection begins at sunrise when the oil yield is at its highest, and should be completed by 10.00 am whilst the dew is still on the flowers. The flowers are initially placed into baskets, and then transferred to sacks for transportation to the distilleries.

Time is of the Essence

Whilst the harvesters are picking the flowers, other workers carefully transfer the flowers from the baskets to the transportation sacks where they are weighed, and all the relevant details are recorded since harvesters are paid by the weight of flowers picked. Each sack weighs approximately 25 kilos when full and is loaded onto horse drawn carriages, the backs of donkeys or less commonly, trucks!

The harvest is then transported to the distillery as quickly as possible, since the picked flowers will begin to deteriorate immediately as precious volatile oil begins to evaporate due to the heat of the sun. This in turn of course will lower the yield of the crop and push up the price of production.

Extraction

In Bulgaria during the early 1900's, virtually all rose oil was distilled on-site using direct-fire stills operated by the farmers. A suitable site would be chosen adjacent to the field and near a stream and the apparatus would be set up. Although this sounds rather primitive, the yield produced from this type of amounts to 1 kilo of oil for every 2,500 to 3,000 kilos of roses. Amazingly, this is a considerably higher figure than can be achieved by modern industrial distillation techniques! Modern stills are made of copper and are heated with an open wood fire from below. The roses can not be distilled in the usual way by directly injecting steam, because the petals compact to form a large mass that the steam can not penetrate. Therefore the distillation techniques have been refined in various ways to overcome this problem.

During distillation a large amount of oil is absorbed into the distillation water, and this is known as the 'First Water'. The rose oil must be recovered from this water to produce an acceptable yield, and this is achieved by skilfully re-distilling the water to separate the oil; a process known as cohobation.

The amount of oil produced directly from distillation is as low as only 20% or 25%, the majority being recovered from the distillate water by cohobation. This ratio does vary depending upon certain factors, but is usually in the region of 25% 'direct oil' and 75% 'water oil'. The 'Second Water' remaining after the process of cohobation is then sold as rose hydrosol (aka floral water) or recycled in the still for the next batch of flowers.

The total yield of oil will depend upon several conditions; climate, the time of the harvest, condition of the flowers and the method of distillation. During the middle of the harvest period the yield is higher than at the beginning, and mild weather will result in a further increase in the oil produced. On average, Rosa damascena will yield 1 kilo of oil per 4,000 kilos of flowers using modern distillation processes. Under very favourable conditions only 2,600 kilos of roses may be required to produce 1 kilo of oil, whereas under less favourable conditions up to 8,000 kilos of flowers may be required to produce the same amount of oil.

Aromatherapy Bases

An advanced range of hypoallergenic, unfragranced base products enriched with the therapeutic, beautifying and revitalising qualities of botanical extracts to provide a single, high performance aromatherapy treatment for your clients. The Quinessence Aroma-botanicals Base Collection is an advanced aromatherapy delivery system, specially formulated to unite the benefits of your essential oils with the beautifying and revitalising qualities of our natural, botanically enriched bases. Your aroma - our botanicals.

Pure and natural botanicals of Aloe Vera, Comfrey, Ginseng, Goldenrod, Shea Butter and Seaweed are used extensively to nourish, tone and beautify skin. At the same time, these revolutionary bases work as a traditional carrier to deliver your essential oils in one easy to use, high-performance treatment.

As a practicing clinical aromatherapist myself, I truly appreciate that the products used in your clinic must produce outstanding results for your clients, whilst also being safe on delicate or sensitive skin. Therefore, when developing this range I insisted on using only natural, vegetal ingredients that have a long and proven record of safety and efficacy. We purposely avoided the use of any modern controversial ingredients that can irritate sensitive skins. Although 'miracle' ingredients such as AHA's and fruit acids promise great results, most of them have not been in general use long enough to discover any problems caused by long-term use on the skin. Every one of the botanical ingredients in this range has been used safely and effectively for centuries, and has been scientifically proven to possess highly beneficial effects.

When these active botanicals are combined with effective vegetable oils such as Centella asiatica, Jojoba, Rosehip and Carrot, they offer a huge range of skincare benefits. Of course, without essential oils it would not be true aromatherapy skincare, so we have ensured that our bases will easily absorb the addition of your essential oils to guarantee truly outstanding results.

Product stability is a crucial factor that must be considered when choosing which supplier you will be entrusting your reputation with. Adding essential oils or vegetable oils to an aromatherapy base that has not been specifically designed for this purpose can result in the product eventually destabilising and breaking down. Over time, the oil and water will separate out resulting in an unusable product, an unhappy client and possibly your reputation damaged.

Aroma-Botanicals are guaranteed to accommodate the addition of your essential oils or vegetable oils without causing separation or disturbing the integrity of the product. Once blended, you can rest assured your aromatherapeutic treatments will remain both highly effective and stable. To ensure suitability for the most sensitive of skin types, these light textured, non-greasy products do not contain any animal extracts, lanolin or harsh chemical preservatives, and are of course fragrance-free, hypo-allergenic and non-comedogenic.

And finally, as a long standing BUAV (British Union for the Abolition of Vivisection) approved manufacturer, I can assure you that every single product in the Quinessence Aromatherapy Collection is manufactured without causing any suffering to animals.

Aromatherapy

Aromatherapy, commonly associated with complementary and alternative medicine (CAM), is the use of volatile liquid plant materials, known as essential oils (EOs), and other scented compounds from plants for the purpose of affecting a person's mood or health. Aromatherapy is a generic term that refers to any of the various traditions that make use of essential oils sometimes in combination with other alternative medical practices and spiritual beliefs. It has a particularly Western currency and persuasion. Medical treatment involving aromatic scents may exist outside of the West, but may or may not be intended by the term 'aromatherapy'.

History: Aromatherapy has roots in antiquity with the use of aromatic oils. However, as currently defined, aromatherapy involves the use of distilled plant volatiles, a twentieth century innovation. The word, aromatherapy, was first used in the 1920s by French chemist Rene Maurice Gattefosse, who devoted his life to researching the healing properties of essential oils after a lucky accident in his perfume laboratory. In the accident, he lit his arm on fire and thrust it into the nearest cold liquid, which happened to be a vat of lavender oil. Immediately he noticed surprising pain relief, and instead of requiring the extended healing process he had experienced during recovery from previous burns—which caused redness, heat, inflammation, blisters, and scarring—this burn healed remarkably quickly, with minimal discomfort and no scarring.

The main branches of aromatherapy include:

- Home aromatherapy (self treatment, perfume & cosmetic use)
- Clinical aromatherapy (as part of pharmacology and pharmacotherapy)

- Aromachology (the psychology of odors and their effects on the mind)

Materials: Some of the materials employed include:

- *Essential Oils:* Fragrant oils extracted from plants chiefly through distillation (*e.g.* eucalyptus oil) or expression (grapefruit oil). However, the term is also occasionally used to describe fragrant oils extracted from plant material by any solvent extraction.
- *Absolutes:* Fragrant oils extracted primarily from flowers or delicate plant tissues through solvent or supercritical fluid extraction (*e.g.* rose absolute). The term is also used to describe oils extracted from fragrant butters, concretes, and enfleurage pommades using ethanol.
- *Phytoncides:* Various volatile organic compounds from plants that kill microbes. Many terpene-based fragrant oils and sulfuric compounds from plants in the genus "Allium" are Phytoncides, though the latter are likely less commonly used in aromatherapy due to their disagreeable smells.
- *Herbal Distillates or Hydrosols:* The aqueous by-products of the distillation process (*e.g.* rose water). There are many herbs that are used to make herbal distillates and they have culinary uses, medicinal uses and skin care uses. Common herbal distillates are rose, lemon balm and chamomile.
- *Infusions:* Aqueous extracts of various plant material (*e.g.* infusion of chamomile)
- *Carrier Oils:* Typically oily plant base triacylglycerides that are used to dilute essential oils for use on the skin (*e.g.* sweet almond oil)

Theory: When aromatherapy is used for the treatment or prevention of disease, a precise knowledge of the bioactivity and synergy of the essential oils used, knowledge of the dosage and duration of application, as well as, naturally, a medical diagnosis, are required.

In the English-speaking world, practitioners tend to emphasize the use of oils in massage. In the UK, America and Australia, aromatherapy tends to be regarded as a complementary modality at best and a pseudoscience at worst.

On the continent, especially in France, where it originated, aromatherapy is incorporated into mainstream medicine. There, the use of the antiseptic, antiviral antifungal and antibacterial properties of oils in the control of infections is emphasized over the more "touchy feely" approaches familiar to English speakers. In France some essential oils are regulated as prescription drugs, and thus administered by a physician. French doctors use a technique called the aromatogram to guide their decision on which essential oil to use. First the doctor cultures a sample of infected tissue or secretion from the patient. Next the growing culture is divided among petri dishes supplied with agar. Each petri dish is inoculated with a different essential oil to determine which have the most activity against the target strain of microorganism. The antiseptic activity manifests as a pattern of inhibited growth.

In many countries essential oils are included in the national pharmacopeia, but up to the present moment aromatherapy as science has never been recognized as a valid branch of medicine in the United States, Russia, Germany, or Japan.

Essential oils, phytoncides and other natural VOCs work in different ways. At the scent level they activate the limbic system and emotional centres of the brain. When applied to the skin (commonly in form of "massage oils" *i.e.* 1-10% solutions of EO in carrier oil) they activate thermal receptors, and kill microbes and fungi. Internal application of essential oil preparations (mainly in pharmacological drugs; generally not recommended for home use apart from dilution - 1-5% in fats or mineral oils, or hydrosoles) may stimulate the immune system.

Choice and purchase: Oils with standarized content of components (marked FCC, for Food Chemical Codex) have to contain X amount of certain aroma chemicals that normally occur in the oil. But there is no law that the chemicals cannot be added in synthetic form in order to meet the criteria established by the FCC for that oil.

For instance, lemongrass essential oil has to contain 75% aldehyde to meet the FCC profile for that oil, but that aldehyde can come from a chemical refinery instead of from lemongrass. To say that FCC oils are "food grade" then makes them seem natural when in fact they are not necessarily so.

Undiluted essential oils suitable for aromatherapy are termed therapeutic grade, but in countries where the industry is not regulated, therapeutic grade is based on industry consensus and is not a regulatory category. Some aromatherapists take advantage of this situation to make misleading claims about the origin and even content of the oils they use. Likewise, claims that an oil's purity is vetted by mass spectrometer or gas chromatography have limited value, since all such testing can do is show that various chemicals occur in the oil.

Many of the chemicals that occur naturally in essential oils are manufactured by the perfume industry and are used to adulterate essential oils because they are cheaper. There is no way to distinguish between these synthetic additives and the naturally occurring chemicals. The best instrument for determining whether an essential oil is adulterated is an educated nose. Many people can distinguish between natural and synthetic scents, but it takes experience.

Price: Oils vary in price based on the amount of the harvest, the country of origin, the type of extraction used (steam distillation, CO_2 extract, enfleurage), and how desirable the oil is. Indian Sandalwood (Santalum album) is considered more desirable than Australian Sandalwood (Santalum spicatum), based upon the aroma, and is twice as costly, mainly because the species that yields Indian Sandalwood essential oils is endangered. Organic and wild harvested essential oils also tend to be more expensive.

Popular uses:

- Basil is used in perfumery for its clear, sweet and mildly spicy aroma. In aromatherapy, it is used for sharpening concentration, for its uplifting effect on depression, and to relieve headaches and migraines. Basil oil has many chemotypes and some are known to be emmenagogues and should be avoided during pregnancy.
- Bergamot is one of the most popular oils in perfumery. It is an excellent insect repellent and may be helpful for both the urinary tract and for the digestive tract. It is useful for skin conditions linked to stress, such as cold sores and chicken pox, especially when combined with eucalyptus oil. Bergamot is a flavoring agent in Earl Grey tea. But cold-pressed Bergamot oil contains bergaptene, a strong photosensitizer when applied to the skin, so only distilled or 'bergaptene-free' types can be topically used.
- Black pepper has a sharp and spicy aroma. Common uses include stimulating the circulation and for muscular aches and pains. Skin application is useful for bruises, since it stimulates the circulation.
- Citronella oil, obtained from a relative of lemongrass, is used as an insect repellant and in perfumery.
- Tea tree oil and many other essential oils have topical (external) antimicrobial (*i.e.* antibacterial, antifungal, antiviral, or antiparasitic) activity and are used as antiseptics and disinfectants.

- Eucalyptus oil
- Sandalwood oil
- Thyme oil
- Clove oil is a topical analgesic, especially useful in dentistry. It is also used an antiseptic, antispasmodic, carminative, and antiemetic.
- Lavender oil is used as an antiseptic, to soothe minor cuts and burns, to calm and relax, and to soothe headaches and migraines.
- Yarrow oil is used to reduce joint inflammation and relieve cold and influenza symptoms.
- Jasmine, Rose, Sandalwood and Ylang-ylang oil are used as aphrodisiacs.

Criticism: The consensus of the position of medical professionals in the U.S.A. and England is that while pleasant scents can be relaxing, lowering stress and offering related effects, there is insufficient scientific proof of the effectiveness of aromatherapy. Scientific research on the cause and effect of aromatherapy is limited, although in-vitro testing has revealed some antibacterial and antiviral effects. Some benefits that have been linked to aromatherapy, such as relaxation and clarity of mind, are quite subjective and may arise from the placebo effect. Like many alternative therapies, few controlled, double-blind studies have been carried out-a common explanation is that there is little incentive to do so if the results of the studies are not patentable. Customers should be aware that aromatherapy may be unregulated, depending on the country. There are some treatments generally accepted in Western medicine to give a form of relief for the airways in case of cold or flu, such as mint and eucalyptus essential oils.

Skeptical literature suggests that aromatherapy is based on the anecdotal evidence of its benefits rather than proof that aromatherapy can cure diseases. Scientists and medical professionals acknowledge that aromatherapy has limited scientific support but argue that its claims go beyond the data or that the studies are neither adequately controlled nor peer reviewed. If there can be positive effects, there can also be negative ones if used incorrectly or in bad combinations, especially with traditional pharmacology. Most medical professionals are concerned that people with maladies curable by contemporary medicine will revert to certain holistic medicines, such as aromatherapy, homeopathy and Ayurvedic medicine, and receive no benefit while their health could have been maintained with scientifically proven medicine. The term "aromatherapy" has been applied to such a wide range of products that almost anything which contains essential oils is likely to be called an "aromatherapy product", rendering the term somewhat meaningless in that context.

Some proponents of aromatherapy believe that the claimed effect of each type of oil is not caused by the chemicals in the oil interacting with the senses, but that the oil contains a distillation of the "life force" of the plant from which it is derived that will "balance the energies" of the body and promote healing or well-being by purging negative vibrations from the body's energy field.

Arguing that there is no scientific evidence that healing can be achieved, and that the claimed "energies" even exist, many skeptics reject this form of aromatherapy as pseudoscience or even quackery. In addition, there are potential safety concerns.

Since essential oils are so potent, many can irritate the skin and can cause toxic reactions like liver damage and seizures unless diluted with a carrier oil such as sweet almond oil, olive oil, hazelnut oil, and rosehip seed oil. Phototoxic reactions may occur with certain citrus oils such as lemon or lime .

Essential Oil

An essential oil is any concentrated, hydrophobic liquid containing volatile aroma compounds from plants. They are also known as volatile or ethereal oils, or simply as the "oil of" the plant material from which they were extracted, such as oil of clove. The term essential indicates that the oil carries distinctive scent of the plant, not that it is an especially important or fundamental substance. Essential oils do not as a group need to have any specific chemical properties in common, beyond conveying characteristic fragrances. They are not to be confused with essential fatty acids.

Essential oils are generally extracted by distillation. Other processes include expression, or solvent extraction. They are used in perfumes and cosmetics, for flavoring food and drink, and for scenting incense and household cleaning products.

Various essential oils have been used medicinally at different periods in history. Medical applications proposed by those who sell medicinal oils vary from skin treatments to remedies for cancer, and are often based on historical use of these oils for these purposes. Such claims are now subject to regulation in most countries, and have grown correspondingly more vague, to stay within these regulations.

Interest in essential oils has revived in recent decades, with the popularity of aromatherapy, a branch of alternative medicine which claims that the specific aromas carried by essential oils have curative effects. Oils are volatilized or diluted in a carrier oil and used in massage, or burned as incense, for example.

Production

Distillation: Today, most common essential oils, such as lavender, peppermint, and eucalyptus, are distilled. Raw plant material, consisting of the flowers, leaves, wood, bark, roots, seeds, or peel, is put into an alembic (distillation apparatus) over water, As the water is heated the steam passes through the plant material, vaporizing the volatile compounds. The vapours flow through a coil where they condense back to liquid, which is then collected in the receiving vessel. Most oils are distilled in a single process. One exception is Ylang-ylang (Cananga odorata), which takes 22 hours to complete through a fractional distillation. The water recondensed from the distillation process is referred to as a hydrosol, hydrolat, herbal distillate or plant water essence, which may be sold as another fragrant product. Popular hydrosols are rose water, lavender water, lemon balm, clary sage and orange blossom water. The use of herbal distillates in cosmetics is increasing. Some plant hydrosols have unpleasant smells and are therefore not sold.

Expression: Most citrus peel oils are usually expressed mechanically, or cold-pressed. Due to the large quantities of oil in citrus peel and the relatively low cost to grow and harvest the raw materials, citrus-fruit oils are cheaper than most other essential oils. Lemon or sweet orange oils that are obtained as by-products of the commercial citrus industry are even cheaper.

Prior to the discovery of distillation, essential oils (EO) were extracted by pressing.

Solvent Extraction: Most flowers contain very little volatile oil to undergo expression and their chemical components are too delicate and easily denatured by the high heat used in steam distillation. Instead, a solvent such as hexane or supercritical carbon dioxide is used to extract the oils. Extracts from hexane and other hydrophobic solvent are called concretes, which is mixture of essential oil, waxes, resins, and other lipophilic (oil soluble) plant material. Although highly fragrant, concretes contain large quantities of non-fragrant waxes and resins. As such another solvent, often ethyl alcohol, which only dissolves the fragrant low-molecular weight compounds, is used to extract the fragrant oil from the concrete. The alcohol is removed by a second distillation, leaving behind the absolute.

Supercritical carbon dioxide is used as a solvent in supercritical fluid extraction. This method has many benefits, including avoiding petrochemical residues in the product. It does not yield an absolute directly. The supercritical carbon dioxide will extract both the waxes and the essential oils that make up the concrete. Subsequent processing with liquid carbon dioxide, achieved in the same extractor by merely lowering the extraction temperature, will separate the waxes from the essential oils.

This lower temperature process prevents the decomposition and denaturing of compounds and provides for a superior product. When the extraction is complete, the pressure is reduced to ambient and the carbon dioxide reverts back to a gas, leaving no residue. Although supercritical carbon dioxide is also used for making decaffeinated coffee, the actual process is different.

Production Quantities: Estimates of total production of essential oils are difficult to obtain. One estimate, compiled from data in 1989, 1990 and 1994 from various sources gives the following total production, in tonnes, of essential oils for which more than 1,000 tonnes were produced.

Oil	*Tonnes*
Sweet orange	12,000
Mentha arvensis	4,800
Peppermint	3,200
Cedarwood	2,600
Lemon	2,300
Eucalyptus globulus	2,070
Litsea cubeba	2,000
Clove (leaf)	2,000
Spearmint	1,300

Essential Oil Use in Aromatherapy

Aromatherapy is a form of alternative medicine, in which healing effects are ascribed to the aromatic compounds in essential oils and other plant extracts. Many common essential oils have medicinal properties that have been applied in folk medicine since ancient times and are still widely used today. For example, many essential oils have antiseptic properties, though some are stronger than others.. In addition, many are claimed to have an uplifting effect on the mind, though different essential oils have different properties. The claims are supported in some studies and unconfirmed in others.

Dilution: Essential oils are usually lipophilic (literally: "oil-loving") compounds that usually are not miscible with water. Instead, they can be diluted in solvents like pure 100% ethanol (alcohol), polyethylene glycol, or oils.

Raw Materials: Essential oils are derived from various parts of plants. Some, like orange oil, are derived from any of several parts of the plant.

Berries

- Allspice
- Juniper

Seeds

- Almond
- Anise
- Celery
- Cumin
- Nutmeg oil

Bark

- Cassia
- Cinnamon
- Sassafras

Wood

- Camphor
- Cedar
- Rosewood
- Sandalwood

Rhizome

- Ginger

Leaves

- Basil
- Bay leaf
- Cinnamon
- Common sage

- Eucalyptus
- Lemon grass
- Melaleuca
- Oregano
- Patchouli
- Peppermint
- Pine
- Rosemary
- Spearmint
- Tea tree
- Thyme
- Wintergreen

Resin

- Frankincense
- Myrrh

Flowers

- Chamomile
- Clary sage
- Clove
- Geranium
- Hyssop
- Jasmine
- Lavender
- Manuka
- Marjoram
- Orange
- Rose
- Ylang-ylang

Peel

- Bergamot
- Grapefruit
- Lemon
- Lime
- Orange
- Tangerine

Root

- Valerian

Rose Oil: The most well-known essential oil is probably rose oil, produced from the petals of Rosa damascena and Rosa centifolia. Steam-distilled rose oil is known as "rose otto" while the solvent extracted product is known as "rose absolute".

Dangers: Because of their concentrated nature, EO's generally should not be applied directly to the skin in their undiluted or "neat" form. Some can cause severe irritation or provoke an allergic reaction. Instead, essential oils should be blended with a vegetable carrier oil (also referred to as a base or "fixed" oil) before being applied. Common carrier oils include olive, almond, hazelnut and grapeseed. Common ratio of essential oil disbursed in a carrier oil is 0.5-3% (most less than 10%) and depends on its purpose. Some EO's including many of the citrus peel oils, are photosensitizers, increasing the skin's reaction to sunlight and making it more likely to burn. Industrial users of essential oils should consult the material safety data sheets (MSDS) to determine the hazards and handling requirements of particular oils.

Pesticide Residues

There is some concern about pesticide residues in EO's, particularly those used therapeutically. For this reason, many practitioners of aromatherapy choose to buy organically produced oils.

Ingestion: While some advocate the ingestion of essential oils for therapeutic purposes, this should never be done except under the supervision of a professional who is licensed to prescribe such treatment. Some very common EO's such as Eucalyptus are extremely toxic internally. Pharmacopoeia standards for medicinal oils should be heeded. EO's should always be kept out of the reach of children. Some oils can be toxic to some domestic animals, cats in particular. Owners must ensure that their pets do not come into contact with potentially harmful essential oils.

Smoke : The smoke from burning essential oils may contain potential carcinogens, such as polycyclic aromatic hydrocarbons (PAHs). Essential oils are naturally high in volatile organic compounds (VOCs). The internal use of essential oils should be fully avoided during pregnancy without consulting with a licensed professional, as some can be abortifacients in dose 0.5-10 ml.

Media: In 2006, the German movie Perfume: The Story of a Murderer was made on the subject of essential oils. The story takes place in France in the 1700's.

Extraction (fragrance)

Fragrance extraction refers to the extraction of aromatic compounds from raw materials, using methods such as distillation, solvent extraction, expression, or enfleurage. The results of the extracts are either essential oils, absolutes, concretes, or butters, depending on the amount of waxes in the extracted product.

To a certain extent, all of these techniques tend to distort the odour of the aromatic compounds obtained from the raw materials. Heat, chemical solvents, or exposure to oxygen in the extraction process denature the aromatic compounds, either changing their odour character or rendering them odourless.

Maceration/Solvent Extraction

Certain plant materials contain too little volatile oil to undergo expression, or their chemical components are too delicate and easily denatured by the high heat used in steam distillation. Instead, the oils are extracted using their solvent properties.

Organic Solvent Extraction

Organic solvent extraction is the most common and most economically important technique for extracting aromatics in the modern perfume industry. Raw materials are submerged and agitated in a solvent that can dissolve the desired aromatic compounds. Commonly used solvents for maceration/solvent extraction include hexane, and dimethyl ether.

In organic solvent extraction, aromatic compounds as well as other hydrophobic soluble substances such as wax and pigments are also obtained. The extract is subjected to vacuum processing, which removes the solvent for re-use. The process can lasts anywhere from hours to months. Fragrant compounds for woody and fibrous plant materials are often obtained in this matter as are all aromatics from animal sources.

The technique can also be used to extract odorants that are too volatile for distillation or easily denatured by heat. The remaining waxy mass is known as a concrete, which is mixture of essential oil, waxes, resins, and other lipophilic (oil soluble) plant material, since these solvents effectively remove all hydrophobic compounds in the raw material. The solvent is then removed by a lower temperature distillation process and reclaimed for re-use. Although highly fragrant, concretes are too viscous - even solid—at room temperature to be useful. This is due to the presence of high-molecular-weight, non-fragrant waxes and resins. Another solvent, often ethyl alcohol, which only dissolves the fragrant low-molecular weight compounds, must be used to extract the fragrant oil from the concrete. The alcohol is removed by a second distillation, leaving behind the absolute. These types of essential oils, from plants such as jasmine and rose, are called absolutes.

Due to the low temperatures in this process, the absolute may be more faithful to the original scent of the raw material, which is subjected to high heat during the distillation process.

Supercritical Fluid Extraction

Supercritical fluid extraction is a relatively new technique for extracting fragrant compounds from a raw material, which often employs Supercritical CO_2 as the extraction solvent. When carbon dioxide is put under high pressure at slightly above room temperature, a supercritical fluid forms (Under normal pressure CO_2 changes directly from a solid to a gas in a process known as sublimation.) Since CO_2 in a non-polar compound has low surface tension and wets easily, it can be used to extract the typically hydrophobic aromatics from the plant material. This process is identical to one of the techniques for making decaffeinated coffee.

Due to the low heat of process and the relatively unreactive solvent used in the extraction, the fragrant compounds derived often closely resemble the original odour of the raw material. Like solvent extraction, the CO_2 extraction takes place at a low temperature, extracts a wide range of compounds, and leaves the aromatics unaltered by heat, rendering an essence more faithful to the original.

Since CO_2 is gas at normal atmospheric pressure, it also leaves no trace of itself in the final product, thus allowing one to get the absolute directly without having to deal with a concrete. It is a low-temperature process, and the solvents are easily removed.

In supercritical fluid extraction, high pressure carbon dioxide gas (up to 100 atm.) is used as a solvent.

Ethanol Extraction

Ethanol extraction is a type of solvent extraction used to extract fragrant compounds directly from dry raw materials, as well as the impure oils or concrete resulting from organic solvent extraction, expression, or enfluerage. Ethanol extracts from dry materials are called tinctures, while ethanol washes for purifying oils and concretes are called absolutes.

The impure substances or oils are mixed with ethanol, which is less hydrophobic [than the solutes?] and dissolves more of the oxydized aromatic constituents (alcohols, aldehydess, etc.), leaving behind the wax, fats, and other generally hydrophobic substances. The alcohol is evaporated under low-pressure, leaving behind absolute. The absolute may be further processed to remove any impurities that are still present from the solvent extraction. Ethanol extraction is not used to extract fragrance from fresh plant materials; these contain large quantities of water, which would also be extracted into the ethanol.

Distillation: Distillation is a common technique for obtaining aromatic compounds from plants, such as orange blossoms and roses. The raw material is heated and the fragrant compounds are re-collected through condensation of the distilled vapour.

Today, most common essential oils, such as lavender, peppermint, and eucalyptus, are distilled. Raw plant material, consisting of the flowers, leaves, wood, bark, roots, seeds, or peel, is put into an alembic (distillation apparatus) over water.

Steam Distillation: Steam from boiling water is passed through the raw material for 60-105 minutes, which drives out most of their volatile fragrant compounds. The condensate from distillation, which contain both water and the aromatics, is settled in a Florentine flask. This allows for the easy separation of the fragrant oils from the water.

The water collected from the condensate, which retains some of the fragrant compounds and oils from the raw material. This fragrant water is called hydrosol and is sometimes sold for consumer and commercial use. This method is most commonly used for fresh plant materials such as flowers, leaves, and stems. Popular hydrosols are rose water, lavender water, and orange blossom water. Many plant hydrosols have unpleasant smells and are therefore not sold. Most oils are distilled in a single process. One exception is Ylang-ylang (Cananga odorata), which takes 22 hours to complete distillation. It is fractionally distilled, producing several grades (Ylang-Ylang "extra", I, II, III and "complete," in which the distillation is run from start to finish with no interruption).

Dry/destructive Distillation: The raw materials are directly heated in a still without a carrier solvent such as water. Fragrant compounds that are released from the raw material by the high heat often undergo anhydrous pyrolysis, which results in the formation of different fragrant compounds, and thus different fragrant notes. This method is used to obtain fragrant compounds from fossil amber and fragrant woods where an intentional "burned" or "toasted" odour is desired.

Expression

Expression as a method of frangrance extraction where raw materials are pressed, squeezed or compressed and the oils are collected. In contemporary times, the only fragrant oils obtained using this method are the peels of fruits in the citrus family. This is due to the large quantity of oil is present in the peels of these fruits as to make this extraction method economically feasible. Citrus peel oils are expressed mechanically, or cold-pressed. Due to the large quantities of oil in citrus peel and the relatively low cost to grow and harvest the raw materials, citrus-fruit oils are cheaper than most other essential oils. Lemon or sweet orange oils that are obtained as by-products of the commercial citrus industry are among the cheapest citrus oils.

Expression was mainly used prior to the discovery of distillation, and this is still the case in cultures such as Egypt. Traditional Egyptian practice involves pressing the plant material, then burying it in unglazed ceramic vessels in the desert for a period of months to drive out water. The water has a smaller molecular size, so it diffuses through the ceramic vessels, while the larger essential oils do not. The lotus oil in Tutankhamen's tomb, which retained its scent after 3000 years sealed in alabaster vessels, was pressed in this manner.

Enfleurage

Enfleurage is a two-step process during which the odour of aromatic materials is absorbed into wax or fat, then extracted with alcohol. Extraction by enfleurage was commonly used when distillation was not possible because some fragrant compounds denature through high heat. This technique is not commonly used in modern industry, due to both its prohibitive cost and the existence of more efficient and effective extraction methods.

Carrot Seed Oil

Carrot seed oil is the essential oil extract of the seed from the carrot plant. The oil is steam distilled from the dried fruit. The odour is a dry-woody, earthy sweet smell, a yellow or amber-coloured to pale orange-brown liquid. Carrot seed oil can be found in different formulas dealing with skin conditions. The oil assists in removing toxins and water build up, giving the skin a fresher tone.

In Aromatherapy

The light yellow to amber coloured essential oil distilled from the seeds (mostly in France, Egypt, and India) differs from the orange carrot oil produced from the root of the common edible carrot. Carrot seed oil nonetheless offers the same familiar scent. Diuretic and hepatic, carrot seed operates as a kidney and liver cleanser, particularly indicated for jaundice and hepatitis. Its depurative (detoxifying) properties are likewise effective for treating arthritis and rheumatism. As a diuretic remedy for genito-urinary ailments, carrot seed is indicated for gout, cystitis, and calculi as well as edema.

Carrot seed stimulates lymphatic circulation and is otherwise carminative and vermifuge, and an emmenagogic menstrual regulator. Its relaxing qualities are helpful in the treatment of premenstual tension. It is also considered a blood tonic indicated for anemia. As a dermal agent, carrot seed lends its depurative qualities to skin and body care preparations and procedures. A natural tanning agent and skin toner that protects aged and wrinkled skin, carrot seed is also remedial for dermatitis, eczema, psoriasis, and various rashes. It can be used topically to treat boils, abscesses, and skin ulcers.

Anethole

Anethole (or trans-anethole) is an aromatic compound that accounts for the distinctive "licorice" flavor of anise, fennel, and star anise. It may also be referred to as p-propenylanisole, anise camphor, isoestragole, or oil of aniseed. It is unrelated to glycyrrhizic acid, the compound which makes licorice taste sweet. The full chemical name is trans-1-methoxy-4-(prop-1-enyl)benzene. The chemical structure is shown at right. Chemically, it is an aromatic, unsaturated ether.

Anethole appears as white crystals at room temperature. Its melting point is 21 °C, and its boiling point is 234 °C. It has a chemical formula of C10H12O, and is closely related to estragole, an aromatic compound found in tarragon and basil. Anethole is distinctly sweet as well as having its flavoring properties and is measured to be 13 times sweeter than sugar. It is perceived as being pleasant to the taste even at higher concentrations. It is slightly toxic and may act as an irritant in large quantities. It can stimulate hepatic regeneration in rats, and can also produce spasmolytic activity in high doses. It is a chemical precursor for paramethoxyamphetamine (PMA), which has been sold as ecstasy resulting in several deaths.

Eucalyptol

Eucalyptol is a natural organic compound which is a colourless liquid. It is a cyclic ether and a monoterpene. Eucalyptol is also known by a variety of synonyms: 1,8-cineol, limonene oxide, cajeputol, 1,8-

epoxy-p-menthane, 1,8-oxido-p-menthane, eucalyptol, eucalyptole, 1,3,3-trimethyl-2-oxabicyclo[2,2,2]octane, cineol, cineole.

Composition: Eucalyptol comprises up to 90 percent of the essential oil of some species of eucalyptus (*e.g.* Eucalyptus polybractea), hence the common name of the compound. It is also found in bay leaves, mugwort, sweet basil, wormwood, rosemary, sage and other aromatic plant foliage. Eucalyptol with a purity from 99.6 to 99.8 percent can be obtained in large quantities by fractional distillation of eucalyptus oil.

Health Warning: In common with all volatile oils (essential oils), eucalyptus oil is toxic if ingested internally.

Properties: Eucalyptol has a fresh camphor-like smell and a spicy, cooling taste. It is insoluble in water, but miscible with ether, ethanol and chloroform. The boiling point is 176 °C and the flash point is 49 °C.

Uses: Because of its pleasant spicy aroma and taste, eucalyptol is used in flavourings, fragrances, and cosmetics. It is also an ingredient in many brands of mouthwash and cough suppressant. Eucalyptol has been demonstrated to be capable of reducing inflammation and pain. It has also been found to be able to kill leukaemic cells.

Cineol was shown to be an effective treatment for Non-purulent sinusitis in a placebo controlled trial. Laryngoscope. 2004 Apr;114(4):738-42. PMID: 15064633 [PubMed - indexed for MEDLINE] 76 patients per treatment group were assigned to cineole or placebo. The dosage of the active ingredient was two 100-mg capsules of cineole three times daily. Symptom scores were significantly reduced in the cineole group. The mean values for the symptoms-sum-scores in the cineole group were 6.9 +/- 2.9 after 4 days and 3.0 +/- 2.8 after 7 days, and in the placebo group, 12.2 +/- 2.5 after 4 days and 9.2 +/- 3.0 after 7 days. Treated subjects experienced less headache on bending, frontal headache, sensitivity of pressure points of trigeminal nerve, impairment of general condition, nasal obstruction, and rhinological secretion. Side effects from treatment were minimal.

In a 1994 report released by five top cigarette companies, eucalyptol was listed as one of the 599 additives to cigarettes. It is added to improve the flavor.

Essential and Organic Oils

Amrette Seed: These have warmth- giving, relaxing and stimulating properties. 2-3 drops of oil are used to relieve anxiety, depression, tiredness and other stress related conditions, cramps, pains and muscular aches. Can freshen up tired and hurting feet. Should be used in low dosage and frequent massages and bath with it should be avoided.

Aniseed: Have warming and stimulating properties. 3-5 drops are helpful in treating bronchitis, coughs and catarrh.

Angelica: Angelica is useful in muscular pains and aches, rheumatism, flu, cold, cough and aides digestion. Used in healing smoker's cough. Not to be used during pregnancy and avoid exposure to sunlight.

Basil: Have uplifting and refreshing properties. Has multiple uses 2-3 drops help to relieve tension, stress, mild anxiety, loss of appetite, indigestion, flatulence, nausea, temporarily relieves cough, sinusitis, cold, fever, bronchitis, earaches, eases rheumatic, arthritic, muscular pains and spasm. It also used as an insect repellent. But is to be avoided during pregnancy.

Bergamot: (3-5 drops) Has calming, refreshing and rejuvenating properties. Can heal wounds, relieves ulcers, treats pre menstrual tension and eczema. Acts as an appetizer. Relieves respiratory infections, colic, flatulence and indigestion. Used in oily skin problems. Used in treating mental and psychological disturbances. It is effective in treating cold sores and chicken pox. Acts as an antiseptic for urinary tract problem. Can be photosensitive and can cause skin irritation.

Black Pepper: Can be warming and strengthening. 2-3 drops can relieve stiffness, pains, muscular aches and fatigue. Helps in curing heartburn, indigestion, flatulence, colic and loss of appetite. Tones muscles and helps in peripheral circulation. Increases concentration and alertness. Provides temporary relief from respiratory infections, chills, cold, flu, arthritis, fibrosis and catarrh. But should used in moderation.

Camphor: Should be used in moderation. Helps in treating colds, coughs, fever, constipation, arthritis, rheumatism, sprains, insomnia, acne, inflamed skin and depression. 2-3 drops can be used at a time.

Cardamom: 2-3 drops can be used. It is a tonic and helps to relieve aches, headaches, coughs, nausea and indigestion. But should be used in moderation.

Chamomile : (3-5) drops can give a soothing effect. Used for diarrhea, gastritis, colitis, urinary tract infection, menopause and menstrual related problems. Eases fear, anxiety and anger. It is helpful in treating sleeplessness and helps to relieve muscular and joint pains and is used in treating skin problems. Soothes inflamed wounds, broken capillaries, blisters. It can be used to lighten fair hair. Has mild effect on children facing teething problem and earache.

Clove Bud: 1-2 drops when used have warming and soothing properties. It is used as an effective antiseptic, analgesic and anti-bacterial. It provides relief from sprains, asthma, cold, flu, dyspepsia, diarrhea, toothache, ulcers, wounds, bronchitis, nervous tension, depression and stress. It is useful in fighting acne and acts as mosquito repellent. It should be used in low moderation and is to be avoided during pregnancy.

Cypress: 2 drops of this can provide refreshed, calm and strengthening effect. It brings down external bleeding, excessive perspiration and heavy menstrual flow; used as an astringent to circulation, hemorrhoids, varicose veins; useful in asthma, bronchitis, dry cough, cold and flu. Helps to heal swollen breasts. It should not be used during pregnancy.

Clarysage : Clarysage has Euphoric, sensual, warming and relaxing properties. 2-5 drops can be used. Brings down high blood pressure and excessive sebum. It is useful in dealing with anxiety, stress, tension and migraine, menstrual cramps and pain, PMT, muscular tension and excessive perspiration. It encourages labour and eases childbirth. Immune system is strengthened and reduces asthma. Causes drowsiness, so it has to be avoided if one suffers from epilepsy or in case of pregnancy.

Eucalyptus: 2-3 drops of this have cooling and clearing properties. It cures cold and fever. Clears blocked nose and head. Relieves stress, fatigue, muscular pains, sprains, aches, rheumatism, headaches, sinusitis and mild respiratory infections. It has anti-inflammatory and antibacterial properties too. And is also used as an insect repellent. Sometimes it can cause skin irritation.

Fennel have calming effect. 2-3 drops are useful in easing colic, constipation, nausea, flatulence, loss of appetite, fluid retention, menstrual pain, PMT, menopause and digestion problems. Has mild laxative and diuretic effect. Cures kidney stones and relieves intestinal spasm. Increases flow of breast milk, relieves bronchitis and helps to fight hangovers. It is to be avoided during pregnancy.

Frankincense it has strengthening and soothing properties. 3-5 drops can be used at a time. Helps in meditation. Strengthens emotionally against grief, fear and nightmares. It cures asthma, bronchitis, coughs and catarrh. It helps in healing dry, wrinkled, scarred, inflamed skin, ulcers and infected wounds. It helps to maintain supple skin.

Geranium : It is associated with calm, balancing, relaxing and refreshing properties. 5-7 drops helps to deal menopausal problems, postnatal depression, PMT. Normalizes skin imbalances and problems. Treats swollen and painful breasts, fluid retention problems, throat infections. It cures cold, flu, acne, bruises, burns, cuts, mouth ulcers, eczema and dermatitis. It acts as astringent and cleanser. Also works as insect repellent.

Ginger: 2-3 drops have warm and stimulating properties. Helps in blood circulation, aids memory, relaxes blood vessels. If inhaled, relieves morning sickness. Relieves pains, aches, sprains and strains. Cures cold, catarrh, flu, fever and sore throat. Relieves flatulence, travel sickness, indigestion. It may cause irritation in sensitive skin so use in moderation.

Juniper Berry: 2-3 drops can be used to heal bruises, acne, colic, cramps, menstrual pain, cough, arthritis, less menstrual flow, swollen and painful breasts. Tones the skin and deals with obesity, loss of appetite, hemorrhoids, poor circulation and fluid retention. If used in excess can irritate kidneys. Not be used if suffering from high B.P. and in pregnancy.

Lavender: It has harmonizing, soothing, balancing, refreshing, relaxing and calming properties. 5-10 drops relieves muscular aches and pains, bites and stings. Cures cold, flu, sleeplessness, headache and minor burns. It helps in relieving anxiety, stress and irritability. It helps to balance and harmonize the body. Usage should be avoided during pregnancy.

Lemon: 3-6 drops help in healing cold, fever, flu, sore throat, mouth ulcers, poor circulation, high B.P., hormonal headaches, wounds, stress, nervous tension, acidity and bronchial problems. Helps to stop hot flushes during menopause. Helps to clear freckles and is an effective insect repellent too. Avoid using it in sunlight.

Lemongrass: (1-3 drops) It helps to open the skin pores, treating oily skin, curing acne and indigestion. Helps in shock, grief and stress. Helps to fight lethargy, insomnia and irritability during menopause, excitability in children. Improves skin elasticity, scars and stretch marks. It gives relief from sore throat, headaches and mild respiratory problems. Can cause skin irritation. To be avoided during pregnancy.

Marjoram: (3-5 drops) Calming to the nervous system. Heals bruises and reduces sexual urges. Relieves anxiety, stress, high B.P., cold, flu, sinusitis, arthritis, rheumatism and swollen joints. Cures indigestion, constipation, flatulence, migraines, stomach and menstrual cramps. Use in limit and avoid during pregnancy.

Neroli: (1-3 drops) Helps in shocks, trauma, grief, stress, menopause and PMT. Cures dryness, improves elasticity of the skin. Calms restless children. Helps to lighten stretch marks and scars. Helps during pregnancy and labour.

Orange (sweet): (2-3 drops) Gentle on children and brings down fever. Eases constipation and diarrhea. Treats nervous tension, stress and anxiety, effective in healing mouth ulcers. Gives relief from bronchial coughs, colds, flu and insomnia. Avoid using under sunlight.

Palmarosa: (4-5 drops) Relieves cold, flu, indigestion, viruses and reduces fever. Helpful in treating acne, scars, dermatitis, maintains supple skin. Stimulates circulation, cell regeneration and appetite.

Peppermint: (1-3 drops) Cools skin inflammation, burns and sunburns. It is a decongestant and gives relief from cold, fever, flu, asthma, headaches, travel sickness, sinusitis, nausea, toothaches, stomach upsets, indigestion and hangover. Avoid during early pregnancy and on sensitive skin.

Patchouli: (2-4 drops) Helps to heal chapped rough skin, inflamed skin, wounds and sores. It is effective in treating dandruff and scars. Helps to relieve anxiety, stress and depression. Used in perfumery due to its exotic aroma. It is used as insect repellent. Excess dose should be avoided as it can cause sedation.

Petit grain: (4-5 drops) It builds resistance against illness. Helps in stimulating digestion and poor memory. Reduces excess acne and sweating, treats greasy hair and skin. It aids in reliving mental strain, anxiety, stress, flatulence, dyspepsia, restlessness and constipation.

Pine: (1-3 drops) It heals cuts and abrasions, excessive perspiration. Relieves colds, coughs, flu, sore throats, asthma, bronchitis, sinusitis, catarrh, muscular aches, arthritis, rheumatism, poor circulation, stress, fatigue, sciatica and poor concentration. Also used to repel insects.

Rose: (1-4 drops) Useful for dry, chapped, aging skin and eczema. Relieves asthma, dry cough, nausea, shock, poor circulation and palpitations. Used to treat irregular heavy menstruation, menstrual pain, PMT, irritability, impotence. Also helps in easing depression, tension, headache, frigidity and insomnia.

Rosemary: (2-4 drops) Improves concentration while studying. It clears dandruff and lice, adds luster to hair. Useful in relieving stress, anxiety, muscular aches and pains, rheumatism, arthritis, sinusitis, migraine, cold, flu, asthma, respiratory problem, poor circulation, menstrual cramps and scanty flow. Avoid usage if suffering from B.P. or epilepsy and during pregnancy.

Sandalwood: (2-5 drops) Heals dry, chapped, cracked, inflamed, mature, aging skin, acne, scars and blemishes. Acts as a deodorizer. Excites the senses. Helps to relieve dry coughs, sore throat, bronchitis, stress, depression, tension, anxiety, insomnia, and travel sickness. It is also an effective insect repellent.

Tarragon: (2-3 drops) It is anti-spasmodic, antiseptic and slightly diuretic. Useful in treating stomach disorders, indigestion, constipation, flatulence, cramps, PMT and nervousness.

Tea Tree: (2-5 drops) helpful in treating colds, flu, cold sores, shock, burns, bacterial and viral infections, nappy rash, stings, herpes, hysteria and in healing fungal and yeast infections.

Thyme (White) : (2 drops) A good stimulant and expectorant. Helpful in relieving cold, coughs, tension, anxiety, fatigue, skin irritation, headaches, rheumatic aches and pains. Also an useful insect repellent.

Vetiver: (1-3 drops) Heals cuts, wounds, blemishes and acne. Helpful in relieving stress, sprains, tension, muscular aches, pains, arthritis, rheumatism, stiffness, palpitations and congestion.

Ylang Ylang: (2-3 drops) Useful in skin care, high B.P., insomnia, PMT, menopause, impotence, frigidity, scalp conditioner. Helps to relieve stress, anxiety, tension, uncontrolled anger, rapid breathing and heart rate. If taken in excess it may cause headache and nausea. Not to be used before driving.

Essential Blends

Aromatherapy is one of the best natural therapies used to rejuvenate mind and body. It makes use essential oils - oils extracted from plants and herbs. It is a combination of two words "aroma" which means scent and "therapy" which means treatment. Aromatherapy was discovered by a French scientist in 1930.

The aroma of the essential oils directly affects the mental and physical health of the patient. Aromatherapy oil also helps in creating emotional balance in emotionally disturbed people. Some of the essential oils are eucalyptus, cedarwood, sandalwood, rosemary etc. Essential oils are highly volatile and therefore should be stored in an airtight container. Essential oils should be mixed with the career oil before using.

Every essential oil has its own unique application and is very effective when used for that purpose. How ever one could also blend 2 or more essential oils to get greater effect than those oils would give individually, when essential oils are blended in this way they are known as essential blends. Essential blends are the undiluted essential oils blended carefully to have a certain effect on mind and body. Essential oils must be blended cautiously. One must keep in mind for what purpose the oils are being blended.

For example, if one is blending oils for diabetes oils like angelica root should be avoided as they are not good for diabetes patient. For blending only natural products as such as essential oils, grain alcohol, carrier oils, herbs and water are used. Basically essential oils can be categorized in to 9 basic categories, Floral Woodsy, Earthy, Herbaceous, Minty, Spicy, Oriental and Citrus. Generally oils in the same category blend well together. Care must be taken that contradictory essential oils are not mixed to create an essential blend *e.g.* an energizing essential oil should not be blended with a relaxing essential oil.

Essential Organic Aromatherapy Oils

Synergies are the blends of organic aromatherapy oils. They are formed by mixing the various organic aromatherapy oils in different proportions or by mixing these oils with the right carrier oils. The power of each oil is magnified when blended together. These synergies can be added to your baths, inhaled, massaged or vaporized for getting the therapeutic benefits of the organic aromatherapy oils. These synergies have beautiful aromas and these smells combined with the soothing effects of the massages and baths produce the relaxation that you have wanted and thus promoting good health by the means of stress reduction.

- *After Flight Synergy:* This synergy uses the blend of geranium, ginger and ylang ylang. Drop a few drops on your tissue or sprinkle them in your bath or shower to eliminate the after flight tiredness.
- *Anti Pollen Synergy:* if you are allergic to the pollen, use the synergy of eucalyptus radiate, lemon and chamomile roman. You get the best results by inhaling the drops or using them in a vapouriser. If you are unable to do so, you can add these drops to your bath or mixed with a carrier oil for massage.
- *Antiseptic Synergy*—You can get the antiseptic synergy by blending together the organic aromatherapy oils of tea tree, lavender and manuka. This is an excellent first aid kit.

- *Anti-virus Synergy*—The combination of ravensara, tea tree and cedar Virginian has excellent anti-viral properties and hence are ideal for use during the colds and any other viral infections. This blend is warming, soothing and clearing to the nose and throat and can be used in bath, inhaled or vaporized.
- *Breathe Easy Synergy*—A blend of cypress, pine and frankincense is refreshing and clearing for you after travelling through the polluted streets. Inhale a few drops on a tissue or vaporize it for immediate relief.
- *Cellulite Synergy*—Tone the skin and increase circulation by blending juniper berry, sweet fennel and cypress. It helps the body break down the fatty deposits. Add a few drops to your bath or massage it on your skin.
- *Energizing Synergy*—Blend grapefruit, pine and litsea cubeba to energize yourself after a tiring day. Recharge your batteries while driving by sprinkling a cotton ball with a few drops of this blend.
- *Foot Ease Synergy*—Blend peppermint, benzoin and patchouli to revive your feet after a tiring day.
- *Head Ease Synergy*—Inhale or vapourize the blend of lavender, peppermint and chamomile roman to ease the head tensions.
- *Immune Optimize Synergy*—Geranium, tea tree and lemon when blended together give a powerful boost to your immune system. Add a small amount to your bath, vapourise it or add to a carrier oil for the massage.
- *Joint Mobility Synergy*—Keep your joins smooth and healthy by using a synergy of ginger, juniper berry and chamomile german. Add a few drops to the bath or to the carrier oil for massage.
- *Menopause Synergy*—Use the blend of geranium, clary sage and cypress for relief.
- *Muscle Ease Synergy*—The blend of marjoram sweet, rosemary and lavender when added to a carrier oil for massage or to the bath water will provide immediate relief to your tired, aching muscles.
- *Pre-menstrual Synergy*—If you suffer from pre-menstrual symptoms, use the blend of geranium, marjoram sweet and chamomile german. Add this blend to your bath or to a carrier oil for massaging your abdomen during these stressful feminine times.

- *Problem Skin Synergy*—Use the blend of lavender, bergamot and chamomile roman to cure red, dry, itchy skin. You can add this blend to any lotion before use.
- *Relaxing Synergy*—Soothe your nerves by using a blend of ylang ylang, chamomile roman and neroli which can be added to the bath.
- *Restful Sleep Synergy*—The blend of bergamot, clary sage and sandalwood when added to the bath will give you a peaceful sleep as this blend contains the most relaxing oils known in aromatherapy.
- *Sinus Synergy*—Keep your sinuses clear and healthy by using the blend of silver fir, eucalyptus radiate and basil. Inhale the drops from a tissue or use it in vaporizer.
- *Stress Synergy*—Use the blend of basil, juniper and geranium to soothe and relax the mind and body. Basil is an excellent tonic for the mind to give it strength and clarity.
- *Uplifting Synergy*—An elevating blend of grapefruit, bergamot and geranium will lift your spirits if you are depressed. It aids in concentration while driving.

Carrier Oils

Carrier oils are commonly known as base oils or vegetable oils. They are used to dilute the essential oils, CO2s and absolutes before being used on the skin or added to aromatherapy bath oils. These oils are called carrier oils because they carry the essential oils to the skin. The properties of different oils are different. You need to choose your carrier oil wisely to get the desired therapeutic effect.

These oils are generally cold-pressed vegetable oils. They are obtained from the fatty portions of the plant like nuts and seeds. While aromatherapy bath oils are highly volatile and fragrant, these carrier oils do not evaporate or release strong smells. Never use the commercially available vegetable oils in aromatherapy.

This is because these oils are obtained by solvent extraction and then destroyed. This process destroys the benefic properties of the oil and thus they become ineffective. Using cold-pressed oil will ensure that the vitamins and therapeutic benefits of the oil are retained during manufacture.

The most common types of carrier oils are sweet almond, apricot kernel, grapeseed, avocado (both refined and unrefined), black seed, borage, calendula, coconut, jojoba, peach kernel, rose hip, St. John's wort, sunflower, peanut, olive, pecan, macadamia nut, sesame, evening primrose, walnut and wheat germ. Depending on the benefit that you are looking out for, choose your oil. These oils by themselves too have therapeutic properties. The carrier oils sold in the grocery store are not cold-pressed but are heated and hence have minimal therapeutic value. These oils should not be used in aromatherapy. Mineral oil should never be used in aromatherapy since it is not a natural product. It prevents the absorption of aromatherapy bath oil into the skin.

Unlike aromatherapy bath oils, carrier oils can go rancid. Always buy the natural and unadulterated oil. But you can buy oils with vitamin E added to it since vitamin E is the most popular natural preservative. The carrier oils should be light and non-sticky which makes their penetration into the skin very easy. Always purchase these oils from a reputed manufacturer. Take references of the experts in aromatherapy for good-quality brands before purchasing any carrier oil.

Almond oil (Sweet) - extracted from the kernel. Contains glycosides, minerals, vitamins and proteins. It is good for skin, helps to relieve itchiness, soreness, dryness and inflammation.

Apricot kernel oil - extracted from the kernel. Contains minerals, vitamins. Used for sensitive, dry, inflamed and all skin types, particularly for pre-matured age.

Avocado pear oil - extracted from the fruit. Contains vitamins, proteins, lecithin and fatty acids. Used in all skin types, especially dehydrated dry skin and eczema.

Carrot oil - contains vitamins, minerals and beta-carotene. Used when there is premature ageing, itching, eczema, dryness and psoriasis. Reduces scars and is very rejuvenating.

Corn oil - contains minerals, vitamins and protein. It is soothing for all types of skin.

Evening Primrose - contains vitamins, minerals and gamma linolenic acid. Used in menopause related tensions, multiple sclerosis, heart diseases. Used in the treatment of eczema and psoriasis. Prevents premature ageing of the skin.

Hazelnut oil - extracted from the kernel, contains vitamins, minerals and proteins. It has a slight astringent action and can be used for all skins.

Peanut oil, Safflower oil, Soya bean oil and Sunflower oil all contain vitamins, minerals and can be used for all types of skin.

Sesame oil - contains vitamins, minerals, protein, lecithin and amino acids and is useful in rheumatism, arthritis, eczema and psoriasis.

Essential Oil Inhalers

"Inhalers" the word brings to ones mind a running or a jammed nose, suffering from cold. To many the word is synonym to any branded product like Vicks inhaler. But actually inhalers are a portable device that gets the medicine directly in to the patient's lungs through breathing. One would be surprised to know that, to many they also serve one of the ways to take aromatherapy.

Aromatherapy heals people using the aroma of natural plants and herbs. Aromatherapy makes use of oils extracted from natural herbs and plants. These oils are known as aromatherapy oil or essential oil. There are many ways of taking aromatherapy like massage, inhaling the aroma, in bath water, dispensers, candles etc.. Inhaling aromatherapy oil affects the central nervous system through the olfactory system. Special inhalers are available in the market for inhaling the essential oil.

Aromatherapy using inhaler is mainly practiced for treating asthma. Essential oil inhalers are portable and could be carried in a pocket or could be hanged around the neck using a string. Using essential inhaler is very simple. Put 8 to 10 drops of desired essential oil or essential blend on a cotton ball and place the cotton ball in to the inhaler. Gently put the essential oil inhaler in one nostril while closing the other and inhale the aroma. Repeat the same procedure for the other nostril too.

Essential oil inhalers are reusable and seal tight and powerful. When the essential oil inhaler is not in use it must be kept inside its container tightly screwed, this will prevent the aroma from leaking out. Using inhalers avoids one from practicing the conventional way of inhaling, where in the patient use to add few drops of essential oil or essential blend to hot water and cover his face with a towel or blanket to inhale the hot vapours. Some of the essential oils or essential blends that could be used with essential oil inhalers are eucalyptus, lavender, chamomile, Rosemary etc. Eucalyptus could be used for treating cough and cold. Lavender could be used for treating insomnia.

As aromatherapy and the essential oils used are completely natural and herbal it does not have any negative side effects.

Hydrosols

The water left behind when you produce essential oil by the means of steam or water distillation is known as hydrosol. It is also called as floral water or distillate water. Since the water soluble parts of the plant are dissolved in this water, hydrosol too retains some of the therapeutic properties of the plant and can be used for healing purpose. Though this water is known as floral water or distillate water, this water can be produced from other parts of plant like herbs, pines, leaves, barks, woods and seeds.

Hydrosol has wonderful aroma of the parent plant. It contains the same therapeutic benefits of the plant. Some plants are specifically distilled for their hydrosol instead of making hydrosol a by-product of the extraction process of the plant. Always ask the seller the details about the hydrosol and try to get some samples before use. This is a crucial since today many hydrosols have synthetic compounds in them. This has caused skin irritation in the sensitive people. Also these compounds do not have any therapeutic properties. Some vendors may also sell water blended with essential oils as floral waters or hydrosols.

Original hydrosols have a wide range of therapeutic values since many active, water-soluble compounds of the plants are present in this water but not in the corresponding essential oil. Hence these floral waters can be used in a wide variety of ways like adding to the bath, sprayed across the room to freshen up your surroundings, to refresh and relax or in skincare products. You can make facial toners and other skin care products. Add them to bath for a refreshing experience or use as a light cologne or body spray. You can add a drop or two in a finger bowl for an elegant, romantic dinner. The common hydrosols are: rose, roman chamomile, chamomile german, peppermint, Melissa, orange blossom (neroli) and lavender.

These hydrosols can be used on the elderly and children or on people with sensitive skin. They can be used undiluted unlike essential oils. However, you should not take them internally. Always keep them away from the eyes.

Cooking and Baking with Pure Essential Oils

Since they are the plant products, essential oils contain practically all the minerals, vitamins and healing properties. They have the healing nutrients, oxygenating molecules, amino acid precursors, coenzyme A factors, trace minerals, enzymes, vitamins, hormones etc. that the parent plant contains.

Also, being highly concentrated, they are more potent and have higher therapeutic powers that the plants or herbs from which they are derived. These oils do not lose their healing properties and oxygen molecules. The oils which are 100% pure essential oils and safe can be used for your daily cooking and home use.

These oils are considered as food items and not as medicines by the government. The chemical structure of these oils is similar to that of our human tissues. This makes them compatible with human protein and the body identifies and absorbs them easily. This makes it easy for you to replace the hydrogenated vegetable oil used in baking of cookies, biscuits etc. with these essential oils.

The following are the tips to cook and bake with pure essential oils safely, while maintaining their healing properties:

- To make a stronger spice oils like basil, cinnamon, marjoram, nutmeg, oregano, or thyme, dip a toothpick in the essential oil and stir it in the recipe after cooking.
- Add 1 or 2 drops of lightly fragrant oil like citrus oils lemon, orange, tangerine before serving to prevent the oil from evaporating to make the recipe for 6-10 people.
- Most of the oils are highly volatile and hence have to be added before serving. Some strong oils like basil, oregano and rosemary when simmered will produce a strong odour.
- Dilute the essential oils with vegetable oil, agave syrup, almond or rice milk before use. Add 1 drop of essential oil to 1 teaspoon of hone, agave syrup or to 2 ounces of beverage.
- Always use the therapeutic quality essential oils.
- Use the cookware you are comfortable with.
- Inhale the aroma of the oil 6 times a day to suppress appetite. It should be inhaled for a longer time. Brief inhaling will have the opposite effect of reversing the appetite. Change the oils daily for the best results.
- Do not use microwave oven as it destroys the enzymes in the food and changes the frequency of the food.
- Avoid using sugar, aspartame and other artificial sweeteners.
- Make ginger cookies with ginger, cinnamon, clove and nutmeg.
- Add lemon, orange, or tangerine oil to the sponge or bundt cake.
- Add peppermint or spearmint oil to chocolate cake, brownie or frosting recipes.
- Improve the taste of the pumpkin pie or spice cake by adding nutmeg, cinnamon, clove or ginger.
- Make tomato sauces, pizza, ravioli, and lasagna recipes more healthy by adding oregano, marjoram, thyme, or basil.

Some common essential oil recipes for cooking:

- o *Salad Dressings or Salad Oils*—Lemon, lavender, rosemary, clove or peppermint in Vegetable Mixing Oil or Massage Oil Base.
- o *Meat and Sauces*—Basil, marjoram, oregano, or thyme.
- o *Cakes, Frosting, Puddings, Druit Pies*—Lemon, clove, orange, tangerine, or peppermint.
- o *Pie Crusts*—Vegetable Mixing Oil produces very flaky crusts.
- o *Herbal Teas*—Lavender, Roman chamomile, orange, tangerine, lemon, peppermint and melissa.
- o *Cool Refreshing Drinks*—Lemon, orange, tangerine, or peppermint added to a pitcher of cold water.
- o *Flavored Honey*—Cinnamon, clove, lavender, basil, chamomile or lemon. (Warm honey until it becomes a thin liquid then add the oil.)

Aromatherapy

The word aromatherapy means "treatment using scents'. It refers to the use of essential oils in Holistic Healing to improve health and emotional well being and in restoring balance to the body. Essential oils are aromatic essences extracted from plants, flowers, trees, fruit, bark, grasses and seeds. There are more than 150 types of oils that can be extracted. These oils have distinctive therapeutic, psychological and physiological properties that improve health and prevent illness. All essential oils have unique healing and valuable antiseptic properties. Some oils are anti-viral, anti-inflammatory, pain-relieving, anti-depressant, stimulating, relaxing, expectorating, support digestion and have diuretic properties too.

Essential oils get absorbed into our body and exert an influence on it. The residue gets dispersed from the body naturally. They can also affect our mind and emotions. They enter the body in three ways: by inhalation, absorption and consumption.

From the chemist's point of view, essential oils are a mixture of organic compounds viz., ketones, terpenes, esters, alcohol, aldehyde and hundreds of other molecules which are extremely difficult to classify, as they are small and complex. The essential oils' molecules are small. They penetrate human skin easily and enter the blood stream directly and finally get flushed out through our elementary system.

A concentrate of essential oils is not greasy; it is more like water in texture and evaporates quickly. Some of them are light liquid insoluble in water and evaporate instantly when exposed to air. It would take 100 kg of lavender to yield 3 kg of lavender oil; one would need 8 million jasmine flowers to yield barely I kg of jasmine oil. Some of these aroma oils are very expensive. They are extracted using maceration. The purification process called defleurage is employed, and in some cases fat is used instead of oil. Then this process, called enfleurage, is used for final purification. Some of the common essential oils used in aromatherapy for their versatile application are:

1. Clary Sage (Salvia Scarea)
2. Eucalyptus (Eucalyptus Globulus)
3. Geranium (Pelargonium Graveolens)
4. Lavender (Lavendula Vera Officinals)
5. Lemon (Citrus Limonem)
6. Peppermint (Mentha Piperita)
7. Petitgrain (Citus Aurantium Leaves)
8. Rosemary (Rosmarinus Officinals)
9. Tea-tree (Melaleuca Alternifolia)
10. Ylang Ylang (Cananga Odorata)

The oils mentioned can be good in a beginner's kit.

Origin of Aromatherapy

The oldest use of aroma oils is known to be as old as 6000 years back when Egyptian physician, Imhotep, the then God of Medicine and Healing recommended fragrant oils for bathing and massaging. In 4,500 B.C. Egyptians used myrrh and cedar wood oils for embalming their dead and 6,500 years later the preserved mummies prove the fact discovered by the modern researchers that the cedar wood contains natural fixative and strong anti-bacterial and antiseptic properties.

Hippocrates, the Greek father of Medicine, recommended regular aromatherapy baths and scented massages. This is what he effectively used to ward off plague from Athens. Romans utilised essential oils for pleasure and to cure pain and also for their popular perfumed baths and massages. Emperor New being indulgent in orgies, feasts and fragrances employed rose frequently to cure his headaches, indigestion and to maintain his high spirits while enjoying amusements.

During the great plague in London in 1665, people burnt bundles of lavender, cedar wood and cypress in the streets and carried posies of the same plants as their only defence to combat infectious diseases. Aromatherapy received a wider acceptance in the early twentieth century. In 1930s Rene- Maurice-Gatte Fosse, a French chemist, diped his burnt hand in lavender oil. To his surprise the wound healed very quickly without any infection or scarring. He did considerable research on various oils and their therapeutic and psychotherapeutic properties.

Dr. Jean Volnet, French army surgeon extensively used essential oils in World War IL It was Madame Morquerite Murry who gave the holistic approach to aroma oils by experimenting with them for individual problems.

Today, research have proved the multiple use of aroma oils. Medical research in the recent years has uncovered the fact that the odours we smell have a significant impact on the way we feel. Smells act directly on the brain like a drug according to scientific research. For instance smelling lavender increases alpha wave frequency in the back of the head and this state is associated with relaxation.

Essential oils like spiritual healing (Reiki, Pranic, Magnified), homeopathic, herbal and flower remedies have a life force that vibrates within the body and the benefit exerted is too subtle to evaluate.

How does Aroma Oils Work?

Dr. Alan Huch, a neurologist, psychiatrist and also the director of Smell and Taste Research Centre in Chicago says, "'Smell acts directly on the brain, like a drug". Our nose has the capacity to distinguish 1,00,000 different smells, (many of which) affect us without our knowing about the same.

The aroma enters our nose and connects with cilia, the fine hair inside the nose lining. The receptors in the cilia are linked to the olfactory bulb which is at the end of the smell tract. The end of the tract is in turn connected to the brain itself. Smells are converted by cilia into electrical impulses that are transmitted to the brain through olfactory system.

All the impulses reach the limbic system. Limbic system is that part of the brain which is associated with our moods, emotions, memory and learning. All the smell that reaches the limbic system has a direct chemical effect on our moods.

For example smelling lavender increases alpha waves in the brain and it is this wave that helps us to relax. A whiff of jasmine increases beta waves in the brain and this wave is associated with an increased agile and alert state. Limbic system is also a storehouse of millions of remembered smells. That is why the mere fragrance of haystack takes us back to childhood. The molecular sizes of the essential oils are very tiny and they can easily penetrate through the skin and get into the blood stream. It takes anything between a few seconds to two hours for the essential oils to enter the skin and within 4 hours the toxins get out of the body through urine, perspiration and excreta.

Aroma oils work like magic for stress-related problems, psychosomatic disorders, skin infections, hair loss, inflammations, pains arising from muscular or skeletal disorders to name some of the application. Actually essential oils have innumerable applications. In Bristol, lavender oil was used on 28 patients who had undergone by-pass surgery. 24 of them reported reduced breathing rates, lower blood pressure and anxiety levels. In Paris, in 1985, 28 women were given treatment for thrush using essential oils. After 90 days the clinical examination showed that 21 of them had been cured completely.

Essential oils are safe to use. The only caution being they should never be used directly because some oils may irritate sensitive skin or cause photo-sensitivity. They should be blended in adequate proportion with the carrier oils. A patch test is necessary to rule out any reactions.

How to Use Aroma Oils?

Essential oils can be used in a variety of ways at home and place of work. Some of the common ways are:

Inhalation: Add 2-3 drops of essential oil depending on which oil you have selected to the hot boiling water and inhale the steam by covering your head with a towel to stop the steam from escaping. Steaming also helps open the pores of the skin and thus more oil is absorbed giving the additional benefit of a facial. The bowl which has the hot water and the aroma oil could be left under the bed so that the room is enveloped in aromatic fragrance. This could be done with the same bowl of hot steaming water and essential oil which had been used earlier for inhalation. A drop or two sprinkled on a handkerchief can give a lasting benefit of the aroma oil. For a very peaceful and relaxed sleep one or two drops of essential oils on a tissue kept inside the pillow or cushion could be used.

Diffusers and Vapourisers: Diffusers are generally made of ceramic or clay. The diffuser has a cave like opening to house small candles or earthen oil lamps and the top is shaped like a curved cup to hold a little water and few drops of aroma oil. Fill the top cup with water add a few drops of essential oils depending on the oil chosen then light the candle or the lamp.

For the oil lamp to last for a long duration, add castor oil to the earthen lamp because castor oil burns for a very long time as compared to the other oils used to light a lamp. Once the water and oil heat up, evaporation takes place and the whole atmosphere is filled with the aromatic scent. The process of evaporation continues for nearly five to six hours. This is ideal for presenting a conducive ambience during a gathering or even in bedrooms, hotels, living -rooms, etc., and anywhere where scented air is required. One can get instant relief from pain a relaxed and positive feeling prevails when the right oil is used. One needs to be careful in choosing the right essential oil.

Vaporisers are the insect repellents used normally in the form of mats or other types of vaporisers kept for repelling insects. One could reuse the used mats by adding 2-3 drops of essential oil of your own choice and keeping them lit (electrically). Slowly the smell will get released and the area would be filled with soothing aroma. Lemon or rosemary are beneficial for offices, lavender for bedroom, antiseptic tea-tree for disinfecting a sick room and citronella for repelling the insects.

Massage: The most common form of treatment is massage because the dual benefits of touch therapy and scent therapy are simultaneously enjoyed. Massage improves the circulation of the blood, tones the muscles, detoxify toxins, releases trapped energy from tense muscles. The fragrance triggers a sense of pleasure and well being. The penetration of essential oil through the skin during massages is high. Generally carrier oils like sunflower, coconut, olive, sweet almond, sesame and vegetable oils are mixed with aroma oils. The aroma oils should not be used for massages directly without dilution. About 10 drops or 1 teaspoon of essential oil can be mixed to about 30 ml of carrier oil. This makes a very rejuvenating massage oil.

Baths: This is an easy way to relax using essential oils.

Foot Bath: You can immerse your feet in a bowl of luke warm water to which 2-3 drops of essential oil is added. This is a very refreshing experience after a hard days work and if you have sweaty and smelly feet then too this foot bath is very profitable.

Pot Pourri: Pot Pourri as the name suggests is a mixture of dried flowers, herbs, grass and seed pods. Few drops of essential oil added to the pot pourri and kept in a bowl would keep giving out aromatic fragrance for 4-6 weeks. Another more effective method would be to keep the pot pourri mixture after adding the essential oil in a closed container overnight so that the oil gets absorbed. The following morning the box can be kept open and the lingering aroma would fill the area.

Bed Time

Sprinkle 2-3 drops on the pillow cover or on a tissue that can be placed under the pillow or cushion cover and inhaled just before sleeping or while sleeping. This can be very useful in treating headaches, stress, tension and in boosting confidence. Some of the essential oils act as an aphrodisiac too.

Compresses

Both cold and hot compresses are profitable. Add 2-3 drops of aroma oil to a bowl of hot (depending on how much of heat you can withstand) or warm water and dip a hand towel or piece of cotton to enable it to absorb the mixture then squeeze out the excess water and place the towel or cotton on the area to be treated. Leaving the compress on the area for 2 hours is quite beneficial. Oil like lavender is usually used. This provides relief when used over bruises, skin problems and pre menstrual syndromes. To make cold compress, add 6 cubes of ice to a bowl with 2-3 drops of essential oil and dip a hand towel or a piece of cotton to absorb the mixture then squeeze out the excess water and place the towel or cotton on the area to be treated. Cold compress is highly helpful in treating burns, sore feet, hangover, sprains and headaches. After a facial the use of hot and cold compress alternately helps the skin.

Oral Intake

It is an accepted practice abroad to take essential oil orally as it is safe. However, care should be taken to take it only under the supervision or guidance of an experienced aromatherapy practitioner. Few oils can be taken internally in prescribed dosage for a particular problem like indigestion under the guidance of a qualified therapist only.

Beauty Treatment

Aroma oils have been used as an application for the skin from times immemorial. As they are highly soothing in treating and enhancing the natural beauty of the skin they can be safely incorporated in facials, massages, manicures, pedicures, scalp treatment, hair wash, hair treatment along with other creams and oils. Rose, chamomile, lemon, lavender, geranium, sandalwood are some good oils for facials irrespective of the fact that beauty treatment is given to normal, mature, dry, oily, sensitive or problem skin. Either one of these or a combination of two of them could be used. The carrier oils that are helpful in a beauty treatment are sweet almond, wheat germ, peach kernel, apricot kernel and sunflower. Steam facials with essential oils are also rejuvenating and help in improving the skin texture.

Room Sprays

There is a call for protecting the environment and this is becoming a prime concern worldwide. Aerosols are being discouraged due to their ozone depleting properties. Essential oils are natural and hence they could be used liberally to deodorise a room, freshen and scent your bathroom, living-room, bedroom, dining-room, office cabin, etc. Merely add 10-12 drops of aroma oil to half a litre of water and spray the mixture with the help of a spray bottle. Oils like lavender, lemon, peppermint, pine and rosemary are best for this application. Cupboards, wardrobes can also be disinfected. If a room smells of dampness or there are moulds in the hotel rooms, houses, offices or factories and shops the essential oil along with water can be sprayed.

Essential Oils and Culinary Herbs

Essential oil plants and culinary herbs include a broad range of plant species that are used for their aromatic value as flavourings in foods and beverages and as fragrances in pharmaceutical and industrial products. Essential oil plants derive from aromatic plants of many genera distributed worldwide. In the United States, the most economically important sources of domestically produced essential oils are industrial by-products from citrus, balsam fir, pine, and cedarwood, while the most important crops grown in the U.S. for essential oils are peppermint and spearmint. Most other essential oils used in the U.S. are imported at an annual cost in 1988 of $150 million (USDA 1989b). Culinary herbs are herbaceous aromatic plants grown and marketed fresh or dried and include many of the same aromatic plants which are grown for their extractable essential oils. Significant quantities of dried culinary herbs are imported annually into the U.S. Recent estimates by the USDA Foreign Agricultural Service reported that more than $349 million of dried condiments, seasonings, and flavourings and $20 million of spice oleoresins were imported into the U.S. in 1988 (USDA 1989a). A significant amount of selected herbs are domestically produced for the dried spice or condiment market Domestic production of these and other herbs and spices now imported is increasing for both processing and fresh market.

The objectives of this paper are to provide an overview to the plants which are processed in the U.S. for essential oils and to identify fresh culinary herbs that are or can be grown in the continental U.S. The potential opportunities and constraints for production of these new crops in American agriculture will be highlighted.

Essential Oils

Chemistry and Extraction of Essential Oils

Essential oils are natural plant products which accumulate in specialized structures such as oil cells, glandular trichomes, and oil or resin ducts. The formation and accumulation of essential oils in plants have been reviewed by Croteau (1986), Guenther (1972) and Runeckles and Mabry (1973). Chemically, the essential oils are primarily composed of mono- and sesquiterpenes and aromatic polypropanoids synthesized via the mevalonic acid pathway for terpenes and the shikimic acid pathway for aromatic polypropanoids.

The essential oils from aromatic plants are for the most part volatile and thus, lend themselves to several methods of extraction such as hydrodistillation, water and steam distillation, direct steam distillation, and solvent extraction (ASTA 1968, Guenther 1972, Heath 1981, Sievers 1928). The specific extraction method employed is dependent upon the plant material to be distilled and the desired end-product. The essential oils which impart the distinctive aromas are complex mixtures of organic constituents, some of which being less stable, may undergo chemical alterations when subjected to high temperatures. In this case, organic solvent extraction is required to ensure no decomposition or changes have occurred which would alter the aroma and fragrance of the end-product. Newer methods of essential oil extraction such as using supercritical CO2 which yield very high quality oils are commercially used, but are less common and beyond the financial means of most processors.

The recovery of nonvolatile essential oils are also obtained by solvent extraction although the process is more difficult and complex than the recovery of the volatiles. This process yields an aromatic resinous product known as an oleoresin, which is more concentrated than an essential oil and which has wide application in the food industry (Heath 1981).

Essential Oils as Industrial By-products

Although a primary focus of this review is to highlight aromatic plants and culinary herbs produced in the U.S., it is important to recognize that the largest quantities of essential oils produced in the U.S. are actually by-products from industrial processes yielding higher value primary products. Citrus essential oils are recovered from the peel which contain the oil sacs or glands located irregularly in the outer mesocarp of the fruit (Matthews and Braddock 1987). These glands are embedded at different depths in the flavedo, the coloured, outer portion of the fruit and must be removed by first rupturing the glands by pressure or mechanical rasping (Matthews and Braddock 1987).

The recovery of citrus oils by mechanical expression is generally obtained by two types of commercial extractors, the FMC Citrus juice Extractor (FMC Corp.) and the Brown Extractor (Automatic Machinery Corp.) (Kealey and Kinsella 1979, Kesterson et al. 1971). Citrus oils are recovered as cold-pressed oils or as a specific constituent such a d-limonene as by-products of the juice and beverage industry and yield important aromatic and flavoring compounds used in a wide array of food, cosmetic and industrial products.

The other large quantity of essential oils produced as industrial by-products in this country comes from the wood and pulp manufacturing industries. More than 1650 tonnes of such oils, predominantly from cedarwood, are produced annually (Lawrence 1979).

How are They Extracted?

Mostly, essential oils are obtained by steam distillation although other methods are used. Citrus fruits are cold pressed by mechanical means, and the oil from delicate flowers is obtained by a more sophisticated method that produces what is known as an absolute. This is because delicate flowers can not withstand the high temperatures needed for steam distillation.After extraction, the resulting essential oil is a highly concentrated liquid that contains the aroma and therapeutic properties of the source from which it came. Nothing should be added or removed from this oil if it is to be used in aromatherapy. To achieve maximum therapeutic benefits, essential oils should be exactly as they came from the still, so to speak.

Standardised Oils

Some industries process essential oils in order to make them meet a required odour or flavour 'profile'. To achieve this, synthetic chemicals are added to the oil and often certain unwanted non-fragrance components are removed (rectification). This 'standardisation' is common practice in the perfumery and flavour industries in order to maintain absolute consistency in fragrance or taste, but totally unacceptable if the essential oil is for use in aromatherapy. To us, this is adulteration - not standardisation. Adulterated essential oils may often smell acceptable to the untrained nose, but because they are extended with synthetics or diluted with vegetable oil it makes them extremely poor value for money. Not only that, but if an essential oil has been standardised, adulterated or adjusted in any way it simply will not be effective.

Look for Purity

This is why you should always buy essential oils from long established and trusted aromatherapy suppliers who specialise in clinical grade oils and not the more common commercial grade. To meet the high quality for aromatherapy, an essential oil should be extracted from a single botanical species that has been botanically authenticated, and derived from a known country of origin. To be 100% pure, nothing should be added or taken away from the oil after extraction. To enable us to meet this requirement, every essential oil supplied by Quinessence has been analytically tested for purity using Gas Chromatography/Mass Spectrometry (GC/MS) to establish its purity. This is why all our oils are guaranteed to be pure, natural and unadulterated.

The Science Bit

The chemistry of an essential oil is extremely complex, and a typical example of oil will contain an elaborate mixture of aromatic constituents such as alcohols, aldehydes, esters, keytones, lactones, phenols, terpenes and sesquiterpenes that combine to produce a unique set of therapeutic qualities. This complex mixture of natural chemicals is what makes essential oils such effective healing agents; for example eucalyptus oil is refreshing and invigorating - plus it is a very powerful antiseptic agent. This means that essential oils can be used in a variety of ways in aromatherapy to promote physical and emotional health and well-being.

Health Benefits

Essential oils possess a wide range of healing properties that can be used effectively to keep you in the best of health as well as looking good. These health-giving benefits include improving the complexion of your skin by stimulating cellular renewal, balancing roller-coaster emotions and fighting bacteria, fungi and other forms of infection. Essential oils have an almost endless list of therapeutic uses. But only by using essential oils of the very highest quality can you be sure of achieving textbook results. Although cheaper essential oils may appear to save you money, they will certainly not deliver the results that you are entitled to expect. Therefore you will have wasted your hard-earned cash. Quinessence are committed to supplying only the very finest, organic essential oils produced from ecologically grown products. We work closely with farmers to ensure that our essential oils are produced from plants that have not been subjected to the use of pesticides or herbicides.

Essential Oil Purity

Using the finest quality essential oils is absolutely vital in aromatherapy. Why? Because the purity of the essential oils will determine how successful the treatment or products will be. In aromatherapy, you only receive textbook results when you use the very purest, highest quality, pure essential oils. At best, essential oils that have been adulterated will simply not deliver the health benefits that you expect. At worst, there is the risk of serious skin irritation or sensitisation due to an adverse reaction to the synthetic chemicals that have been used to adulterate, extend or 'standardise' the essential oil.

Vigorous quality checks are made throughout the entire process of sourcing and authenticating the provenance of Quinessence pure essential oils. Failure to meet any one of our quality standards will result in an essential oil or other raw material being firmly rejected.

Our Quality Control Standards include:

- Verification of the plants botanical species
- Crops were not subjected to agrochemicals
- Low pressure distillation techniques employed
- Visually inspecting the oil
- Odour evaluation of the oil
- Measuring the oils physical parameters
- Testing the oils purity using GC or GC/MS technology

We believe that essential oil quality control should begin at the earliest stage of production, preferably with inspection of the crop at the growing phase. This important aspect is often overlooked when considering quality issues, and this is why we have established close working relationships with growers around the world.

When growing plants specifically for the extraction of essential oils, the location, altitude, fertility of the soil, weather conditions and the method of cultivation all have an influence upon its oil content and quality. This is why an essential oil produced in one country can be of a higher quality than precisely the same oil produced in another. And it doesn't end there. At the extraction stage, the skill and experience of the distillery becomes the next critical factor in the quest for excellence, since all the expertise and effort of the farmer will have been wasted if the oil is not extracted correctly.

The importance of purity with essential oils can not be overstated, if for no other reason than maintaining the good reputation of aromatherapy. Cheaper, inferior quality essential oils deliver poor results and mean that ultimately you will have wasted your hard earned money, rather than saving any of it. Using cheap essential oils can often be a bit of a gamble too. At best, nothing at all will happen - in other words you will not receive any benefits, but at the worst you may have an allergic reaction to the oil and experience skin irritation, sensitisation, or other undesired side effects.

Unwanted side effects are usually caused when an essential oil has been adulterated with a synthetic chemical component, or perhaps because it is entirely synthetic. Unscrupulous traders and dealers will often 'stretch' expensive oils by adding cheap synthetics to them to make the product more profitable for themselves. This type of adulteration can cause problems for unsuspecting customers who then attempt to use these oils in aromatherapy - sometimes with disastrous results. Many people are allergic to synthetic compounds, hence the move to 'fragrance-free' cosmetics. Of course not everyone is sensitive to synthetic or adulterated oils so the problem should not be exaggerated, but it is still unethical and dishonest for a supplier to sell an adulterated or synthetic oil under the pretence that it is a natural, pure essential oil. After all, you are paying for something that you are not getting. Synthetic or adulterated oils contain little or no therapeutic qualities and will not produce the results that you are looking for.

Essential Oil Testing

When it comes to essential oils, purity is everything. And because they are the very heart and lifeblood of all Quinessence aromatherapy products, we move heaven and earth in order to bring you the very best. Part of our quality control procedure is to ensure that every essential oil we purchase has been tested for purity and meets every one of our exacting standards. We make no compromises or exceptions on this matter. Our reputation has been built upon the quality of our essential oils and we remain firmly committed upholding these standards. Therefore, essential oils are never purchased in bulk until they been subjected to the most rigorous qualitative and quantitative analytical procedures to confirm their purity and authenticity. The practice of oil adulteration has now reached a very high level of sophistication, therefore a wide range meticulous checks must be made to ensure that the purity of every single essential oil has been confirmed.

Qualitative and Quantitative Analysis

So what exactly what is the difference between qualitative and quantitative analysis? Put simply, qualitative testing helps to determine exactly what individual constituents are present in an essential oil, and quantitative testing provides information on how much of each component is present. In the hands of experts, this information can be used to verify that the oil is indeed what it claims to be and originates from the claimed country of origin. This is very important because it is well known by experts that for any given essential oil, there are several origins and the quality between them varies tremendously. Therefore, the purchase price varies too.

Sensory Evaluation

The first steps of essential oil testing usually begins with sensory evaluation, since this saves wasting precious time and money on more expensive analytical procedures and can identify inferior oils quickly. The viscosity, colour, clarity and odour of an essential oil can help to identify a poor quality oil right at the outset - provided you know precisely what to look for. For instance, a sample of Rose Otto which appears too mobile at a low temperature is a typical example of an oil which we would immediately regard as very suspicious. Genuine Rose Otto oil congeals at around 16 degrees centigrade because of the natural waxes that are present, therefore if it is still mobile at or below this temperature, it is likely that the oil has been adulterated or 'extended' in some way.

Similarly, Geranium oil can be purchased in bulk from a wide range of geographical locations such as Egypt, China, Africa and Reunion at prices that vary considerably. Since Chinese Geranium oil is green and not a golden colour like the Egyptian material, it isn't difficult to spot the difference visually. It is not uncommon for unscrupulous traders to try and pass off one as the other, since at times of market fluctuations the purchase prices vary considerably.

Odour Evaluation

An odour test can also help to determine if an oil is really what it purports to be, since certain adulterants can be identified in this way. Some essential oils such as Lavender are available from several geographical locations, and a trained nose will be able to detect if a so called 'French' Lavender is in fact from another, less desirable (or less expensive) origin.

Sensory evaluation is a simple but effective method of identifying clumsily adulterated oils quite quickly, but remains only one of many procedures that must be conducted during the search for purity. Other, more sophisticated methods of analysis must be employed before a clearer picture can be formed of an oils authenticity.

Physical Parameters

If an essential oil sample passes all of the sensory tests, the next stage is to test the physical parameters of the essential oil by means of measuring the Specific Gravity, Optical Rotation and Refractive Index. This is a more searching examination that will confirm or reject the authenticity of an oil's declared botanical species and country of origin, whilst possibly revealing any adulteration with a foreign substance. The combination of these physical tests is usually sufficient to determine if it is worth proceeding to the final stage of testing an essential oil. If an oil successfully passes the first two stages it is then tested using Gas Chromatography/Mass Spectrometry (GC/MS).

Gas Chromatography

When using Gas Chromatography to test an essential oil, a tiny sample of the oil is injected (pictured right) into the apparatus which contains a very thin coiled silica tube called a 'capillary column'. This capillary column may measure up to 100 metres in length and is coated on the inside with a material that has an affinity to different chemicals at different temperatures. The column is housed within a temperature regulated oven and is programmed to steadily increase in temperature over a period of time in a very precise manner.

When the sample of oil is injected into the column it immediately vaporises, and an inert carrier gas (usually hydrogen or helium) moves the vapour along the column to a detector called a Flame Ionisation Detector which is situated at the end of the column.

Component Identification

The flame ioniser detector responds quantitatively to the vaporised constituents of the oil and converts this information, via an integrator/ computer, into proportional peaks printed onto computer listing paper. Every individual component of the essential oil can be identified by the time at which the peak elutes on the trace. The data produced can then be compared to an established 'profile' or 'fingerprint' for that particular essential oil to finally determine the purity of the oil.

Adulterants can usually be identified by this means of testing, although it does require the expertise of an organic analytical chemist who is a specialist in this area to fully and accurately interpret the results of the testing. Variations in climatic conditions, and the type of soil in which a plant was grown will produce natural variations in essential oils produced from the same species.

Quality Control

Quinessence is totally committed to supplying you with only the very purest, therapeutically active essential oils and aromatherapy products, manufactured from ecologically grown raw materials and sustainable sources. You want products that really work, and our aim is to exceed your expectation at every level. To achieve this high standard, more than 20 years have been spent building close working relationships with growers, distillers and suppliers around the world. Striving for excellence is what helped us earn our reputation, and ensuring our aromatherapy products continue to meet these very demanding standards is what helps us keep it.

What does This Involve?

Quality control involves processes, equipment and people. And with essential oils it is our opinion that QC should begin at the earliest stage of production, which means inspecting the crop at the growing phase. This important aspect is often overlooked when considering quality issues, and this is precisely why we have insisted on working closely with our main suppliers.

Wherever it is possible, we will visit the farm to inspect their facilities, collect samples and generally get to know the people who we will be doing business with. Even at this early stage we will make checks to verify the botanical species of the plant, and where growers claim to have organic status we insist upon having copies of their documentation to support this. Nothing is left to chance.

Analytical Testing

Before purchase, every essential oil goes through a series of searching analytical procedures to ensure that the oil has not been adulterated, stretched or otherwise 'adjusted' in any way. To achieve this tests are conducted to measure the physical parameters of oils, check the odour profile, and if all is well go for GC/MS analysis. Failure to meet any one of our demanding quality standards - at any stage - will result in an oil or raw material being firmly rejected. Discover more about GC analysis.

Traceability

After passing quality control, the raw material is purchased in bulk and the consignment is given a unique batch number and entered into the records of our raw materials database. Information recorded includes; the product name, product specification, its geographical origin, date of purchase and name of the producer or supplier. This means we are always able to trace every ingredient, of every product, right back to its source.

The same level of meticulous checks are enforced on every one of the raw materials we procure to ensure that our high standards are continually upheld. Compromising on quality is not an option at Quinessence.

Manufacturing Standards

During the process of manufacturing our aromatherapy products, the same exacting standards are rigorously imposed at every stage in order to support full traceability, whilst showing compliance with all relevant legislation and regulations. Products are filled using highly accurate computer-controlled measuring equipment in our 'clean-room' facilities. Visual inspections are made at several stages of filling to confirm that the correct volume of product is being delivered. Further random samples are taken and checked for accuracy prior to the labelling phase. All of our filling equipment is checked regularly by the U.K.'s Trading Standards Authority to ensure that this accuracy is maintained from batch to batch.

Safety

Safety and quality are equally important for aromatherapy products, and we work hard to ensure that our products are suitable for the entire family. All Quinessence products provide full instructions for use, including how many essential oil drops should be used when mixing a massage blend or added them to a bath.

Tamper-evident closures are used on all essential oils, and integral drippers are incorporated to ensure that an unattended child could not swallow the oil. Four different sizes of drippers are used to ensure accurate and easy dispensing of our full range essential oils.

Essential oil product labels provide the botanical name of the oil, its country of origin, instructions for use plus any contraindications that are applicable to the oil in the bottle. These are not just 'generic' instructions - everything is specifically related to the essential oil in the bottle.

Every product has a batch number sticker on the bottle or jar which is recorded on our product database. To help you get the very best performance out of your products, suggested 'best before' dates are also recorded on products wherever applicable. More on this.

Customer Support

We want to provide you with the best Customer Support possible, and therefore apply our same rigorous quality control procedures here too. As a manufacturer and distributor with over 20 years experience we have developed a wide range of systems to deal with the speedy and efficient processing of all incoming and outgoing orders.

A telephone Helpline service is available to all customers, supported by qualified aromatherapists who are members of the International Federation of Aromatherapists (IFA), International Federation of Aromatherapists (IFPA), The English Societe l'Institut Pierre Franchomme, or the Register of Qualified Aromatherapists (RQA). Our senior therapist is a qualified clinical aromatherapist who has over 22 years experience in the aromatherapy industry.

Material Safety Data Sheets (MSDS) are available for professional practitioners (where applicable), and can be obtained on request.

2

Uses for Essential Oils

Aromatherapy/Essential Oils for Headaches

Headaches are among the most common of all medical complaints. At least 90 percent of all Americans have reported having at least one type of headache, with 28 million Americans experiencing migraine on a somewhat regular basis. Cluster headaches are much less common, with less than 0.1 percent of the population reporting this condition. Tension headaches, the most common form of headache, are usually treated with a variety of over-the-counter medications, including analgesics, such as acetaminophen, aspirin, and a variety of other non-steroidal anti-inflammatory drugs (NSAIDs); however, taking these drugs can also cause unwanted side effects, especially if used frequently. Aromatherapy with essential oils can provide an alternative form of treatment to pain medications, offering a natural and safe form of headache relief.

How Can Aromatherapy Be Used to Treat Headache? Aromatherapy is a method of treatment that uses specific essential oils to relieve the symptoms of a disorder. Each essential oil is chosen for treatment because of a specific effect it has on the body. Some of the most popular essential oils used to treat headache include the following:

- Chamomile (*Matricaria chamomilla)* because of its ability to relieve pain and relax the muscles
- Geranium (*Pelargonium odorantissimum*) for its ability to strengthen and invigorate the body
- Lavender (*Lavandula officinalis*) because it is one of the best-known and most-widely used herbs for the relief of pain, its

anti-inflammatory properties, and its ability to soothe and relax the body

- Peppermint (*Mentha piperita*) because of its analgesic and anesthetic properties
- Rose (*Rosa damascena*) because of its sedative and relaxing effects on the nervous system
- Rosemary (*Rosmarinus officinalis*) because it is an analgesic and nervine (a substance that strengthens, calms, and soothes the nerves)

The usual procedure for using essential oils in the treatment of headache is to add a few drops of specific oil to a carrier oil (a neutral oil), such as jojoba oil. The mixture can then be applied to the head, neck, shoulders or other parts of the body affected by the headache by means of massage, compress, or by adding a few drops to a cloth applied to the head. It may help to add a few drops of the mixture to one's pillow before bedtime. The essential oil can also be added to facial oils and applied during routine face care.

Practitioners often recommend a mixture of essential oils that, by combining a variety of properties, can produce a better result. Migraine Relief, one commercially available product, is a blend of six essential oils, aniseed (*Pimpinella anisum*), basil (*Ocimum basilicum*), lavender (*Lavandula officinalis*), marjoram (*Origanum marjorana*), niaouli (*Melaleuca viridiflora*) and peppermint (*Mentha piperita*). The combination is said to have a number of beneficial effects, including calming and soothing the nerves, relieving nervous tension, reducing nausea and indigestion, acting as an antidepressant, alleviating anxiety, and preventing insomnia.

Aromatherapy is a system of healing that makes use of essential oils, organic compounds obtained from the flowers, stems, leaves, roots and other parts of a plant. The oils are removed from the plant parts by steam distillation and are then purified for use. Healers have known for hundreds of years that essential oils have properties that are helpful in treating physical, mental and emotional disorders. These properties include the ability to soothe and relax the body, fight microorganisms that cause disease, reduce inflammation, reduce depression, eliminate body odor, increase the elimination of water from the body, clear the lungs, strengthen the immune system, and improve body functions in many other ways. Aromatherapists are specialists who have been trained in the properties of essential oils and are knowledgeable of the best way they can be used to treat a variety of diseases and disorders.

Headache, also known as cephalagia, is quite literally a pain that occurs in the head, which usually includes the upper neck, areas above the eyes or ears, or in the face. Headaches are considered primary if they are not associated with other diseases or medical conditions, and secondary if they are.

Primary headaches are often classified in one of three ways:

- *Tension headaches*, or myogenic headaches, are the most common type of headaches, are often described as similar to having a tight band around the head causing mild to severe pain that may be focused on the base of the skull or the back of the neck.
- *Migraine headaches*, a form of vascular headaches, are perhaps the most severe type of headache, often accompanied by nausea and vomiting, and photophobia (light sensitivity).
- Cluster headaches, another form of vascular headaches, are also characterised by severe pain that may continue for weeks or months, as enlarged blood vessels press upon nerves in the face.

Headaches have a great variety of causes. Tension headaches may be caused by dietary problems (allergies to certain foods), repetitive physical activities (sitting at a computer for long periods of time), stress and tension (worries about schoolwork or one's job), fatigue, overuse of caffeine or alcohol, smoking, or sleeping in a cold environment. The basic cause of migraine headaches is not known, although they can be precipitated by a number of factors, such as certain foods (such as chocolate, nuts or peanut butter), exposure to bright lights or loud sounds, alcohol consumption, smoking, irregular eating schedules, physical or emotional stress, allergic reactions or changes in one's menstrual cycle. The cause of cluster headaches is also not known. Cluster headaches tend to differ from tension and migraine headaches in some characteristic ways. For example, they tend to begin just before one goes to sleep and produce a number of physical effects (tearing of an eye and stuffiness in one side of the nose) that affects only one side of the face.

Common Cold

A common cold is rightly termed so because of the multiple kinds of colds that an individual goes through in his/her lifetime. A common cold is mostly brought on by the deposition of the rhinovirus (that causes a cold) in the nasal passage of an individual. The common cold is probably the most common type of infection seen across the globe, and children, especially, are more susceptible to this type of viral infection as their immune system is not fully developed.

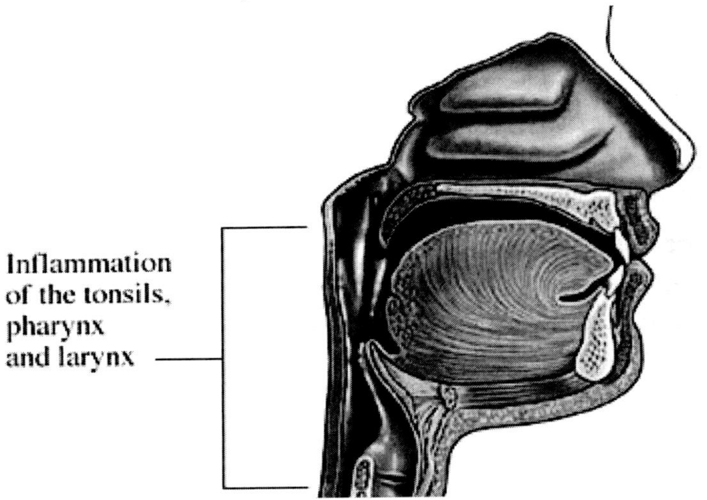

Common colds are often a very mild viral infection that clears up after it has run its course. In isolated incidents, common colds can increase in severity and may lead to advanced infections like bronchitis or pneumonia. Once the cold virus has entered the body of a person, the symptoms of a cold start manifesting in about two to three days. The symptoms peak around the third to fourth day. Then, they wane off in a period of seven to ten days. A common cold is easily picked up at crowded places, schools and childcare centres. The fall and winter months provide the ideal breeding ground for the cold infections to multiply. During these seasons, the cold weather forces people to stay indoors and the spread of the virus is considerable.

Symptoms for Common Cold

The symptoms of a common cold begin within a few days of the incubation period of the cold virus. The symptoms for a common cold that are often seen, in the young and old alike, include:

- A stuffy nose that is accompanied with runny discharge in most cases.
- An itchy and sore throat.
- Change in speech, which becomes quite hoarse and/or nasal
- Watering of the eyes
- Headaches

- Fever that is low grade in nature
- Aches all over the body and in the joints and muscles
- Overall feeling of fatigue.

In children, the heavy headedness and nasal congestion can make them quite irritable, and this adds to their overall discomfort. The symptoms for a common cold generally improve generally in a week's time. Medical attention needs to be sought if one or any of the symptoms cause considerable discomfort to the person and if they persist over two weeks.

Causes for Common Cold

A common cold is a mild viral infection that is extremely ubiquitous in nature. The virus is released into the atmosphere in the form of tiny droplets when an infected person coughs, sneezes, or blows his/her nose. Exposure to cold weather is often cited as one of the causes for a common cold.

- Rhinovirus is one of the most common viruses that can cause a common cold in an individual. There are various strains of the common cold virus, which is why the common cold is so persistent an infection. As your immunity may develop resistance to one strain you could still experience another bout of the common cold when infected by a different strain of the virus.
- The initial days, following the onset of common cold in a person, are the days when the risk of contagion is highest. Close contact with an infected individual puts you at high risk of picking up the infection, particularly if your immunity is low.
- Apart from the virus that causes common cold, people who have a chronic medical condition are also susceptible to colds as their immune system is quite compromised.
- Individuals with low immunity are more prone to catching colds. This is because their external barrier of protection is not as efficient and in such a weakened state the body is a lot more vulnerable to almost any viral or bacterial infection.

When a cold afflicts patients suffering from asthma, diabetes, or heart disease the symptoms of the existing medical condition are greatly aggravated. There may be difficulties in managing the symptoms of both conditions. Thus, the risk of contracting associated complications is higher in such cases.

Remedies for Common Cold

A common cold has to run its course before the individual feels any better. Most over the counter medications for the common cold simply provide temporary relief from the discomfort caused by the symptoms of the infection. As the infection is viral, there is little that medications can do to actually cure the condition. For this reason home remedies for common cold are extremely popular. Natural remedies for a common cold can help to facilitate recovery and provide relief from most symptoms. Because of the recurring nature of coughs and colds most people prefer to rely on home remedies for coughs and colds. Some such remedies include:

Lemon and honey mixed in warm water can be consumed twice a day. The high concentration of vitamin C improves the resistance of the body, lowers the toxic nature of the infection, and reduces the total duration of the cold.

- Cloves of garlic, boiled in water and consumed as a soup, provide relief from the symptoms of a cold. The antiseptic and antispasmodic nature of garlic mitigates the discomfort. The oil contained in the garlic is also beneficial in opening up the airways, thus facilitating better respiration.
- Ginger is also used as one of the home remedies for a cough and cold. Ginger tea, mixed with honey and drunk hot, provides soothing relief to the throat and clears the feeling of heavy headedness.
- Natural cold remedies also include the use of turmeric. The anti-inflammatory properties of this root can help ease the symptoms of a cold and rejuvenate the infected person. It is best to consume a hot glass of turmeric and ginger mixed in milk, before retiring for the night.
- A runny nose is a common symptom associated with a cold. To address a running nose a home remedy banks on the benefits of inhaling the vapours of turmeric, apple cider vinegar, or eucalyptus oil that can be added to steaming water water. This helps loosen the secretions, thus enhancing the discharge through your nose.
- Steam inhalation is an effective home remedy for coughing and headaches that mitigates the intensity of the pain. The headaches associated with a cold are typically because of congestion. The vapours inhaled cause a discharge of the mucus, helping clear congestion.

- Saline gargles are helpful to soothe an itchy and sore throat. The use of humidifiers may also be helpful, particularly if you spend most of your time indoors.

The symptoms of common cold generally require medical attention only when they persist beyond a couple of weeks. The chances for secondary infection need to be ruled out and the physician may suggest further follow-up measures.

Diet for Common Cold

A common cold is best countered with improved strengthened immune mechanism of the body. A generous intake of vitamin C helps to boost your natural defences to fight off infections. Citric fruits, like lemon and orange, are extremely rich sources of vitamin C. A common cold diet should include lots of fruits, fresh, green and leafy vegetables, grains, and nuts. It is best to stick to a light diet rather than one rich in red meats and fatty and processed foods. Chicken soup is generally well tolerated by persons afflicted with a common cold. It helps to clear the nasal blocks and stuffiness. Chicken is also believed to help reduce inflammation. Spinach if added to this is also beneficial as it is a rich source of vitamins A and C. Fish is also recommended in a common cold diet, as the omega-3-fatty acids have an anti-inflammatory effect and soothe the symptoms of a cold.

In addition to these foods, it is essential to increase your intake of fluids. This is because the enhanced hydration keeps the mucous and secretions from causing congestion and helps to dispel them from the body at a faster pace.

The common cold is not just highly contagious, it is also not restricted to any particular age, racial or ethnic group. Anyone can be afflicted with a common cold and due to its highly contagious nature, you need to take precautions against passing on the infection. The best way to reduce the frequency of colds and minimise the risk of spreading infection is to practice good hygiene. If you have been in contact with or the proximity of an individual afflicted with a common cold, make it a point to wash your hands thoroughly and avoid touching your mouth or nose till then. Rest is absolutely essential for a swift recovery, so make sure that you don't tax yourself too much.

A lack of sleep can severely compromise your immunity, leaving you vulnerable to infections and also increasing the duration of the illness. Herbal teas can also be of great help when dealing with a common cold. Simply sip on some hot herbal teas like ginger or peppermint to get relief from the symptoms. This will also help raise your fluid intake and boost immunity. A common cold is only like to turn serious when the affected person is an infant or baby, or aged. This is because of a risk of complications developing.

Using Essential Oils for Arthritis

Arthritis is a disease millions of people suffer from every day. While it's pretty rare that anyone has been able to cure it, arthritis can be treated and controlled. The use of essential oils is one of the best and safest ways to treat and control the pain caused by arthritis. There are two types of pain-acute and chronic-which affect the systems of the body. Acute in general refers to pain that will end ie pain caused by burns, broken bones, headaches, etc. Chronic, on the other hand, refers to pain that is persistent, wearing and gets worse with time rather than better-arthritis, for example. Arthritis attacks the joints, quite often the hands due to their complex arrangement of joints, bones and tendons. When treating joint pain caused by arthritis the first thing to do is look at your diet. Decrease your intake of simple carbohydrates and avoid coffee and red meat. Increase your vegetables, fruits, complex carbohydrates, wheat germ and oily fish.

Massage with essential oils is one of the most effective methods of relief found for arthritis. Some believe arthritis can be cured if caught and treated quickly enough. There is not enough evidence to say whether that is true but the treatment suggested is extremely helpful in providing relief and making it possible to control the level of pain and the amount of activity you can do.

The following is an arthritis bath blend that some have found helpful. For each bath you add 2 handfuls of Epsom salt and one of rock salt, and 4 drops of the following blend:

Fennel	30 drops
Cypress	16 Drops
Juniper	10 Drops

This bath treatment is best used daily for only two weeks. While using it, don't use any other treatments including any other essential oils. (Before using, make sure you are not allergic to any of the oils used in the blend.)

After the two weeks, you can massage any of the following essential oils into the joints that are affected: Roman Chamomile, German Chamomile, Lavender, Yarrow, Peppermint, Niaouli, Calendula,

Eucalyptus (any), Angelica, Cedarwood, Sandalwood, Pettitgraine, Black Pepper, and Rosemary.

Sandalwood is good for smoothing out the skin and when mixed with Lavender has a lovely scent to it as well. Peppermint, Rosemary, Lavender, and Eucalyptus work well as a blend. The Peppermint helps the blood move while also helping to cool down any inflammation.

We also offer a blend called Tension Release which, when combined with Helichrysum and Jojoba Oil and massaged into the affected joints, is wonderfully invigorating and helps alleviate the pain.

When using any essential oils for massage, always add a few drops of whatever oils you use to at least a tablespoon of a carrier oil. Fractionated Coconut Oil (FCO) is one of the best because the skin absorbs it well. Also use a teaspoon of Jojoba Oil in your blend.

Arthritis can be a crippling and extremely painful problem that will last until you die. But it can be managed and, in particular, managed without using drugs. We here at Rocky Mountain Oils provide some of the best Essential Oils on the market which will help you in your efforts to control the effects of this disease so that it won't control you.

Essential Oils for Pain Relief

It is safe to say that a great number of clients come to massage seeking relief from muscle and joint pain. Many will also be suffering from stress and need to relax. Others may be athletic or high-powered performers who want pain relief without becoming tired or drowsy. Essential oils can address all of these areas and add a pleasing fragrant dimension to your therapeutic work.

The most famous essential oil for pain relief and relaxation is lavender (*Lavandula officinalis*, L *angustifolia*, L *vera*.) Distilled from the flowering tops, the best lavender oil comes from Bulgaria, France, England, Yugoslavia and Tasmania, though it can be grown all over the world. Lavender Vera is grown in higher altitudes, which produces more esters and a finer scent. Lavender has a long list of applications for skin; because of its anti-inflammatory and cell regenerating properties, it is one of the only essential oils that can be applied neat, or undiluted, to the skin.

Lavender is also antimicrobial, anti-infectious and antiseptic, making it effective in the treatment of wounds and as a frontline defence against respiratory infection. It is tonic to the cardiovascular and digestive systems, lowers blood pressure and helps thin the blood due to the presence of coumarins. Lavender is indicated for muscle spasm, sprain, strain, cramp, contracture and rheumatic pain. It is sedative to the central nervous system and relieves headache, nervous tension, and insomnia; it can also help balance mood swings. Spiritually, lavender is said to balance the physical, astral and etheric planes.

Because of lavender's many therapeutic properties, if aromatherapists were stranded on a desert island with only one essential oil, many would hope it was lavender (it also takes the itch out of insect bites and helps heal sunburn!) But what other essential oils can be called in to use here in civilization? What should you use if your client does not want the deep relaxation or sleep-inducing effect of lavender, or if they have a tendency toward lowered metabolism or low blood pressure? What if they need to relax because they are about to take an exam, give a presentation or walk down the aisle? It's a good idea to ask the client who indicates a need to relax what their stress is about and what life circumstances may be contributing to their pain cycle. This will help you select an essence that is most appropriate for their needs. Also keep in mind that when too much lavender is used it takes on the stimulating effect of a cup of espresso, so it is good for both you and your client to vary the relaxing, pain-relieving blend.

We'll begin with an exploration of aromatherapy for pain and stress, and profile some other sedative oils. Space allows for a partial list of the properties; consult *The Aromatherapy Practitioner Manual*, Vols. I and II by Sylla Sheppard-Hanger, *Aromatherapy for Healing the Spirit* by Gabriel Mojay, and others for more information on each essence. When you want slightly less sedation but powerful pain relief, there is another type of lavender, *Lavandula latifolia*, L. *spica*, or Spike Lavender. A hybrid of lavender *officinalis* and *latifolia*, Lavandin, Lavandula-super is less expensive and often used to adulterate true lavender but is still a powerful antispasmodic well-suited for muscular, respiratory and circulatory problems, and not as a sedative for the mind.

Moving away from the lavenders altogether, other pain relieving sedative oils are chamomile (Roman, *Anthemis nobilis* and German, *Marticaria recutita*), Clary sage (*Salvia sclarea*), helichrysum (H. *angustifolium*), sweet marjoram (*Origanum majorana*), sandalwood (*Santalum album*) and vetiver (*Vetiveria zizanioides*).

Chamomile is a highly effective anti-inflammatory. It eases headache, neuralgia, dull muscle and low back pain, and TMJ syndrome. It relieves dysmenorrhea, PMS and stress that manifests as digestive symptoms.

Clary sage (not to be confused with sage, *Salvia officinalis*) is considered mildly intoxicating and euphoric, and should be used in small quantities and preferably not before an evening of cocktails, as it augments the effects of alcohol. Apart from this, the ability of Clary sage to relieve spasm, muscle ache and cramping makes it extremely useful in massage. It is a digestive aid and can be blended effectively with chamomile for tension and discomfort due to PMS and dysmennorhea. Along with lavender, Clary sage is one of the essences chosen to ease labour. It is also associated with dreams and increased inner vision. Helichrysum has a long history as anointing oil, but well deserves an honoured place in therapeutic massage. With many of the properties of lavender, helichrysum is also indicated for bruising and burns, depression, shock and phobia, and is helpful in detoxification from drugs and nicotine. Helichrysum is said to improve the flow along the meridians and to increase spiritual awareness.

Sweet marjoram is highly sedative. It relieves pain, stiffness, sprain, spasm, neuromuscular contractions and is indicated for both rheumatoid and osteoarthritis, dysmenorrhea and migraine. It has a powerful effect on the mind and emotions, relieving deep trauma, grief and heartache.

Sandalwood, well known in Ayurvedic treatment and as incense, also relieves muscle spasm and is helpful in treating sciatica and lymph congestion. It is tonic in the cardiovascular and digestive systems and relieves depression, insomnia, obsession, grief and aggression. Sandalwood opens the mind to spiritual connection and grounds this awareness in the material world.

Vetiver is interesting because it relieves arthritis, muscle ache, pain, sprain and stiffness, but increases venous circulation to help detoxification of tissues. It is said to balance the central nervous system and is grounding and revitalising, while relieving insomnia, tension and depression.

Apart from lavender, all of the sedative essences listed are pretty potent and require few drops in a blend. The flower essences: rose, jasmine, neroli and ylang ylang, relieve anxiety and have properties that induce relaxation and pain relief.

The citrus oils: sweet orange, grapefruit, lemon, lime, tangerine and Mandarin, reduce tension and instill courage and optimism. Flower and citrus oils blend well with the other sedative oils and add their own dimensions to the therapeutic experience. If you have a great pain relief or relaxing blend and want to share it, please contact me. In the next Aromatic Message, we'll look at some of the less sedative and stimulating oils for pain relief.

Spices as Aphrodisiacs

The heady aromas of expensive, exotic spices ensured that they would offer a voluptuously stimulating environment for invigoration of romantic encounters. In the Old Testament's the Song of Solomon, Proverbs and Psalms romantic verses extolled the sensory excitement offered by cinnamon, calamus, myrrh, saffron and other perfumed smells from fragrant spices. In Greece and Rome, spices were included in antidotes against poisons and venoms but their potent, life-restoring virtues earned them a heady reputation of being essential every-night aphrodisiacs; indeed, in Rome the word cinnamon was equivalent to the current use of "sweetheart" or "darling". The Romans also embraced the phytochemical concept of the biblical lover's spicy enticement: "Make haste, my beloved, and be thou like to a roe, or to a young hart, upon the mountains of spices." "I have perfumed my bed with myrrh, aloes and cinnamon. Come let us take our fill of love till morning." (Proverbs 7, 17-18). The Arabs had their "Perfumed Garden" and the Hindus their "Kama Sutra", each of which extolled favoured spices such as nutmeg, cloves, galangal, cardamon and ginger, while the Romans came to favour cinnamon and pepper, and the Chinese were most impressed with ginger.

Over the years, spices have offered the luxury of intriguing tastes, impressive incenses and delightful perfumes, and, as tools of the rich, they have always been included in recipes for improving sexual potency. It is of interest that the equivalent of the multi-herb antidote against all poisons that was concocted by King Mithridates VI, who ruled over ancient Turkey, is still on sale in a modern reformulation in that country. It now carries suggestive names such as "Sultan's Paste".

Proprietary luxuries of this type that consist of several dozen herbs and spices are currently promoted as aphrodisiacs and tonics rather than as antidotes against poisoning, or as incenses, for appeasing the gods in religious ceremonies. Undoubtedly, spicy versions of these recipes that served the ancient pagan gods such as Priapus, Cupid, Venus, Eros, Pan and of course Aphrodite (the goddess who arose from sea foam - "aphros") continue to work their historic magic. Modern romances are catalysed by spices and herbs which are called upon to provide symbolic and sensory support in luxurious perfumes, heady scents, and sensual aromatic cream or oil massages. However, it is of interest that the most appreciated of current aphrodisiacs is undoubtedly the New World's Aztec "food of the gods", the meso-American spice chocolate rather than the ancient and historic spices of Arabia and the Orient.

The essential oils and terpenoid alcohols of spices contribute to their smell, taste and tactile sensation. Thus, eugenol is found in cinnamon, clove and pimento; one of its medical qualities is a local anesthetic effect, which is utilised in dentistry. Menthol, from mints, has a cooling effect as well as a characteristic fresh taste and smell. Anise contains anethole, cinnamon produces cinnamaldehyde, mace contains myristin, and so on; all have specific pharmacologic effects that are generally mild. However, some - such as myristicin - are more potent, and large doses can result in harmful effects such as hallucinations.

A number of spice chemicals are shared with herbs and flowers. It is noteworthy that colourful flowers result in an experience of exciting colour and smell, whereas most spices result in excitatory sensations of taste and smell without being particularly stimulating to the visual sense. There are some exceptions, including the crocus which is the source of saffron, and edible flowers such as nasturtium which can spice up a salad. Similarly, chile peppers and radishes can be visually exciting, whereas cinnamon bark and cardamon seeds are relatively dowdy. The following spices have had a long reputation of having aphrodisiacal properties.

- Asafetida This has a foul smell, but in small amounts it can provide a sensual taste or smell. The same phenomenon applies to musk oil (from the musk ox) and castoreum (from the beaver), and perhaps to the secretions of the civet cat and the skunk: these agents can give a salty, animalistic, deeply erotic fragrant quality to a perfume when suitably diluted.
- Cardamon is popular in India and in Arabic cultures, and used to be employed by the Chinese court to give users a fragrant breath.
- Cloves and some other spices and herbs contain eugenol; its smell is fragrant and aromatic, and has long been considered as enhancing sexual feelings.
- Ginger contains gingerols, zingiberene and other characteristic agents that have made it a favoured seductive flavour in Asiatic and Arabic herbal traditions.

- Mace and Nutmeg contain myristicin and similar compounds that are related to mescalin. In larger doses, nutmeg and mace can cause hallucinations, whereas in smaller amounts they are traditional aphrodisiacs.
- Pepper from India contains piperine: this pungent agent can stimulate sexual function, according to ancient beliefs.
- Saffron contains picrocrocin which is alleged to have the ability to cause erotic sensations.
- Vanilla contains the widely loved vanillin, whose taste and smell conjure up romantic feelings in the appropriate circumstances.

Other popular herbs that have been reported to have aphrodisiacal properties include garlic, mint, rosemary, sage and thyme. All these allegedly erotically stimulating agents have long been incorporated into cooking, incenses, rubs and other romantic sources for stimulation of sexual feeling.

More recently, these and other herbs are utilised creatively in numerous massage oils and in incenses that are popularly utilised to improve sensations as a new-old form of therapy, with the modern title of aromatherapy.

50 Uses for Essential Oils

1. For good smelling towels, sheets, clothes, etc. place a few drops of your favourite essential oil onto a small piece of terry cloth and toss into the clothes dryer while drying. Add 5 drops essential oil to 1/4-cup fabric softener or water and place in the centre cup of the washer.
2. Adding a few drops of essential oil can revive potpourri, which has lost its scent.
3. Add a few drops of oil to water in a spray bottle and use as an air freshener.
4. Add a few drops essential oil to a pan of water and simmer on stove or in a potpourri pot.
5. To enjoy a scented candle, place a drop or two into the hot melted wax as the candle bums.
6. To dispel household cooking odours, add a few drops of Clove oil to a simmering pan.
7. For tired aching muscles or arthritis aches, mix 1 part Cinnamon, Sage and Basil oil to 4 parts Sweet Almond or other vegetable oil and use as a massage oil.

8. Ease headache pain by rubbing a drop of Rosemary or Lavender oil onto the back of your neck.
9. To blend your own massage oil, add 3-5 drops of your favourite essential oil to 1 oz. Sweet Almond or other skin- nourishing vegetable oil.
10. Add 10 drops of essential oil to a box of cornstarch or baking soda, mix very well, let set for a day or two and then sprinkle over the carpets on your home. Let set for an hour or more, then vacuum.
11. To make a natural flea collar, saturate a short piece of cord or soften rope with Pennyroyal or Tea Tree oil, roll up in a handkerchief and tie loosely around the animal's neck.
12. Shoes can be freshened by either dropping a few drops of Geranium essential oil directly into the shoes or by placing a cotton ball dabbed with a few drops of Lemon oil into the shoes. Athlete's foot? Tea Tree is great!
13. Put a few drops of your favourite essential oil on a cotton ball and place it in your vacuum cleaner bag. Lemon and Pine are nice. Rose Geranium helps with pet odours.
14. To fragrance your kitchen cabinets and drawers, place a good scent dabbed on a cotton ball into an inconspicuous comer.
15. Are mice a problem? Place several drops of Peppermint oil on a cotton ball and place at problem locations.
16. The bathroom is easily scented by placing oil-scented cotton balls in inconspicuous places, or sprinkle oils directly onto silk or dried flower arrangements or wreaths.
17. Apply true Lavender oil or Tea Tree oil directly to cuts, scrapes or scratches. 1 or 2 drops will promote healing.
18. Homemade soaps are pleasant and offer therapeutic effects when scented with essential oils. Use soaps which contain pure essential oils.
19. Homemade sachets are more fragrant when essential oils are blended with the flowers and herbs.
20. An essential oil dropped onto a radiator scent ring or light bulb will not only fill the room with a wonderful fragrance, but will also set a mood such as calming or uplifting. (Don't put essential oil in the socket.)
21. A few drops of your favourite oil or blend in the rinse water of your hand washables makes for pleasant results.

22. Anise oil has been used by fishermen for years. Use a drop or two on the fingertips before baiting up. Anise covers up the human scent that scares the fish away.
23. Essential oils or blends make wonderful perfumes. Create your own personal essence! Add 25 drops to 1 oz of perfume alcohol and allow to age for two weeks before using.
24. To dispel mosquitoes and other picnic pests, drop a few drops of Citronella oil on the melted wax of a candle or place a few drops on the Bar-B-Q hot coals.
25. 1 drop of Lemon essential oil applied directly to a wart is an effective means of elimination. Apply the essential oil daily until the wart is gone.
26. Rosemary promotes alertness and stimulates memory. Inhale occasionally during long car trips and while reading or studying.
27. Selling your home? Fragrance sells! Fill the kitchen area with the aroma of spices such as Clove, Cinnamon and Vanilla. Simmer a few drops of the essential oil of Cinnamon, Nutmeg and other spices. Geranium oil sprinkled throughout the home creates a warm, cheerful and inviting mood. Add Cinnamon oil to furniture polish and wipe down the wood.
28. Add essential oils to paper mache. The result is the creation of a lovely aromatic piece of art.
29. Infuse bookmarks and stationery with essential oils. Place drops of oil on paper and put them in a plastic bag. Seal it and leave overnight to infuse the aroma. Send only good news in perfumed letters.
30. Nock pillows, padded and decorative hangers make more memorable gifts simply by putting a couple of drops of essential oil on them before giving.
31. Overindulge last night? Essential oils of Juniper, Cedarwood, Grapefruit, Lavender, Carrot, Fennel, Rosemary and Lemon help soften the effects of a hangover. Make your own blend of these oils and use a total of 6-8 drops in a bath.
32. Essential oils of Vetivert, Cypress, Cedarwood, Frankincense and Myrrh all make wonderful firewood oil. Drop approximately 2-3 drops of oil or blend of your choice on a dried log and allow time for the oil to soak in before putting the log on the fire.
33. Flies and moths dislike Lavender oil. Sprinkle it on the outside of your window frames.

34. Place 1 or 2 drops of sleep enhancing oils such as Chamomile, Lavender or Neroli on your pillow before retiring for restful sleep.
35. When moving into a new home, first use a water spray containing your favourite essential oils and change the odorous environment to your own. Do this for several days until it begins to feel like your space.
36. Ideal scents for the bedroom are Roman Chamomile, Geranium, Lavender or Lemon.
37. One drop of Lemon essential oil on a soft cloth will polish copper with a gentle buffing.
38. When washing out the fridge, freezer or oven, add 1 drop of Lemon, Lime, Grapefruit, Bergamot, Tangerine or Orange essential oil to the final rinse water.
39. For bums or scalds, drop Tea Tree oil directly on the effected area.
40. Place 1 drop of Peppermint oil in 1/2 glass of water, sip slowly to aid digestion and relieve upset stomach.
41. Use 1 drop of Chamomile oil on a washcloth wrapped ice cube to relieve teething pain in children.
42. Six to eight drops of Eucalyptus oil in the bath cools the body in summer and protects in winter.
43. Add 1 drop Geranium oil to your facial moisturizer to bring out a radiant glow in your skin.
44. Place 1 or 2 drops of Rosemary on your hair brush before brushing to promote growth and thickness.
45. When the flu is going around add a few drops of Thyme to your diffuser or simmer in a pan on the stove.
46. To bring fever down, sponge the body with cool water to which 1 drop each of Eucalyptus, Peppermint and Lavender oils have been added.
47. The blend of Lavender and Grapefruit oil is good for the office. Lavender creates a calm tranquil atmosphere while Grapefruit stimulates the senses and clears up stale air.
48. A blend of Geranium, Lavender and Bergamot alleviates anxiety and depression. Use in a room diffuser or 6-8 drops of this blend in the bath.
49. A wonderful massage blend for babies is 1 drop Roman Chamomile, 1 drop Lavender, 1 drop Geranium diluted in 2 Tablespoons Sweet Almond oil.

50. 1 drop Peppermint oil diluted in 1 teaspoon vegetable oil rubbed on the back of the neck helps to relieve headaches

Geranium essential oil is antibacterial, antispasmodic, anti-tumorial, adrenal support, anti-inflammatory, astringent, antibacterial, anti-fungal, haemostatic (stops bleeding), and general tonic, Geranium was used by the ancestors as a remedy for wounds, tumours, and skin care. With antiseptic and anti-inflammatory, Geranium essential oil is indicated in acne, dermatitis, eczema, wounds that are weeping and will not heal, bruises, grazes and cuts, light burns, hemorrhoids, varicose veins, and ulcers, sore throats, tonsillitis, and neuralgia.

The essential oil is one of the few that has been used successfully against the MRSA bacteria in laboratory studies (search Pub Med for more on 'geranium oil mrsa'). Geranium oil has also been successfully used in treatment of Candida infections, and has been the subject of many scientific studies validating its antimicrobial actions. Geranium specifically in combination with Lemograss has been studied for its highly effective antibacterial activity, also against antibiotic resistant staph strains.

Often planted around the home for protection, beauty, and aroma, Geranium is a familiar perennial shrub, that grows in stands between three four feet high in herb gardens. The shrub has pointed leaves that are serrated at the edges and hosts delightful pink to white flowers. The name Geranium derives from the Greek word *geranos* or 'crane' because the seed pods resemble the shape of crane's bills.

The plants originated from South Africa, as well as Egypt, Madagascar, and Morocco, and were introduced to European countries such as Italy, Spain and France in the 17th century. Geranium essential oil was first cultivated for use in the French perfume industry for 'rose geranium oil'. Although there are nearly 700 different varieties of Geranium only ten of these supply essential oil in viable quantities and only the scented varieties of Pelargonium are used for the essential oil. Best when steam distilled, Geranium essential oil is a great balancer. A middle note with oil that is clear to light green and has a light fresh scent. Use confidently in blends to balance the heaven and earth qualities of other oils.

Harmonic and balancing, Geranium can be used for spiritual practice and meditation as she assists one in connecting to the intuition and wisdom messages of the heart. This naturally lead to an increased capacity for intimate and open communication between body, mind, and spirit. Geranium essential oil has the quality of equalising hormonal and emotional extremes. Known familiarly as the 'flower of constancy', Geranium helps to lift the spirits and helps to bring joy and happiness to one's daily activity.

Known for its support in releasing negative and dis-harmonious memories, Geranium eases nervous tension and stress and is balancing to the emotions, thus lifting the spirits, and fostering a sense of peace and well-being. Indicated for those who feel the downward spiral of depression and for those whose lives are 'lacking colour', the essential oil is celebrated for its ability to lift worry and anxiety. This clearing and dispelling of stagnant energy can be helpful for those who need the motivation and strength to bring their visions and projects into manifestation. These properties make it an ideal essential oil for stress-related disorders such as headaches, stomach aches, and menstrual cramps.

As both a sedative and uplifting oil, Geranium's action on the nervous system is pronounced. Similar to both Basil and Rosemary, Geranium is a stimulant of the adrenal cortex, whose hormones are essentially of a balancing nature. Consequently it is indicated when the hormonal system needs balancing. This harmonizing effect on the emotions assists in reducing mood swings, and is helpful in cases of PMS and menopausal discomfort.

Connected to the feminine influence of the planet Venus, Geranium has a dynamic flowing energy that helps to balance the water in the body. As a kidney tonic, Geranium assists with releasing fluid retention and in this way is useful for cellulite therapies. Additionally Geranium helps with estrogen production and relieves fluid retention that may occur in some women prior to their moon cycle. Geranium helps regulate estrogen production and relieves edema (fluid retention) that may occur in some women prior to their moon cycle. When used massaged into the breast tissue, it assists in reducing swelling and engorgement thus making it a good remedy for mastitis.

As well as having a balancing effect on the mind, this uplifting essential oil has a great all-over balancing property also extends to the skin, where it helps to create balance between oily and dry skin. An astringent that is not dying, Geranium is soothing to the mucus membranes of the skin. Balancing the sebum production (the fatty acid secretions in the sebaceous glands) of the skin, Geranium helps to cleanse the skin as well as to restore balance, tone, and suppleness.

Other uses include lice and mosquito repellent, and the clearing of ringworms. There is some evidence indicating that Geranium essential oil may be helpful in cases of neuralgia. Geranium's gentle yin characteristics lend it blending well with nearly all other essential oils including eucalyptus, lavender, clary sage, rose, lime, orange frankincense, grapefruit, and ylang ylang. Diffuse and use topically.

Essential oils, or volatile oils, are found in many different plants. These oils are different from fatty oils because they evapourate or volatilise on contact with the air and they possess a pleasant taste and strong aromatic odor. They are readily removed from plant tissues without any change in composition. Essential oils are very complex in their chemical nature. The two main groups are the hydrocarbon terpenes and the oxygenated and sulphured oils.

These oils do not have any obvious physiological significance for the plant. They may represent byproducts or metabolism rather than foods. The characteristic flavour and aroma that they impart are probably to some advantage in attracting insects and other animals, which play a role in pollination or in the dispersal of the fruits and seeds. When in high concentration, these same odours may serve to repel enemies of the plants. The oils may also have some antiseptic and bactericidal value. There is some evidence that they play an even more vital role as hydrogen donors in oxidoreduction reactions, as potential sources of energy, or in affecting transpiration and other physiological processes (Hill 1952).

All the distinctly aromatic plants contain essential oils. They occur in over 60 families and are especially typical of the Lauraceae, Myrtaceae, Umbelliferae, Labiatae and Compositae. The quantity of oil varies from a very small amount to as much as 1-2 percent. The oils are secreted by internal glands or in hair like structures. Sometimes, as in wintergreen and mustard, the oil is not present in the plant but develops only as the result of chemical action when the ground-up plant tissue is extracted with water. Almost any organ of a plant may be the source of the oil. Examples are flowers (rose), leaves (mint), fruits (lemon), bark (cinnamon), wood (cedar), root (ginger) or seeds (cardamom), and many resinous exudations as well. These oils are extracted from the plant tissues in different ways depending on the quantity and stability of the compound. Three principal methods are: expression, distillation and extraction by solvents.

The history of civilization is directly connected with that of perfumes. Perfumes have been in widespread use since the earliest recorded times. The Egyptians and ancient Hebrews used them for both personal and religious purposes. They played an important role in the life of the Romans and Greeks, reaching such a high degree of specialisation with the Greeks that a special perfume was required for each part of the body. Later Catherine de' Medici knew as much about perfumes as she did about poisons. In the time of Queen Elizabeth a gift of rare perfumes was a definite way to win the royal favour, while the court of Louis XIV at Versailles had a particular perfume for each day of the year, the preparation of which was supervised by the king himself. In those days perfumes were of hygienic as well as aesthetic value for they acted as true antiseptics and deodorants and masked offensive odours at a time when bathing were infrequent. Perfumes have continued to be in great demand to the present day. The consumption of the natural products has gradually increased in spite of the many synthetic substitutes that chemists have placed on the market. Synthetics are not as long lasting as those obtained directly from the plants.

The most valuable perfumes are combinations of several essential oils. Frangipani, for example, contains sandalwood, sage, neroli, orris root, and musk, while one of the formulas for Eau de Cologne, which dates from 1709, calls for neroli, rosemary, lemon and bergamot dissolved in pure alcohol and aged. "The expert perfumer must be able to blend the several oils at his command as an orchestra leader combines the various instruments into a perfect whole" (Hill 1952).

Perfumes also contain fixatives, which are substances that are less volatile than the oils and which delay and so equalise evapouration. These may be of plant or animal origin. Musk, ambergris, and civet are frequently used for this purpose. Balsams and oleoresins, such as benzoin, styrax, and oak moss; essential oils with a low rate of evapouration like orris, patchouli, elary sage, and sandalwood; and various synthetic materials are also used.

Perfume plants are cultivated for the most part in areas bordering on the Mediterranean Sea and the Indian Ocean. Most of the natural perfumes are made in southern France in the region around Grasse and Cannes near the French Riviera. Here garden flowers are cultivated on a large scale, and from 10-12 billion pounds were being gathered annually by the mid 1950's.

These included over 5 million pounds of orange blossoms, over 4 million pounds of roses, 440 thousand pounds of jasmine and 330 thousand pounds of violets. Large quantities of tuberoses, cassie, jonquils, thyme, rosemary, lavender and geraniums are grown and many other fragrant species to a lesser degree. Flowers are also grown for the perfume industry to some extent in Reunion, North Africa, England and various European, Pacific and Asiatic areas. When supplies were reduced during World War II, the United States developed substitutes and initiated or increased the cultivation of several essential oil plants in Central America. Of the 75 essential oils regularly used in the industry only eight are normally produced in the Western Hemisphere, and only oil of petitgrain is of much importance.

Perfume Oils

Otto of Roses

This is valuable oil that is also called Attar of Roses. It has been one of the most favourite perfumes either in combination with other oils or alone. Bulgaria supplied most the commercial supply in the 20th Century. The damask rose, *Rosa damascena,* was the main source. By the mid 1900's over 12,000 acres on the southern slopes of the Balkans were devoted to its cultivation. The harvest period covers about three weeks during May-June. Flowers are picked in the early morning just as they are opening and are distilled immediately. In the beginning of this industry peasant farmers utilised their own primitive stills, but this gave way to larger modern distilleries. The oil is colourless at first but gradually turns a yellowish or greenish colour. More than 20,000 lbs of the flowers are required to make one pound of the essence, which was valued at $200.00 in 1952. Very little pure Otto reaches the markets because it is almost always diluted with geranium or palmarosa oil or geraniol, which also have a rose like odor. Otto of Roses is also manufactured in France, Italy, North Africa, Asia Minor and India. In France the cabbage rose, *R. centifolia* was used and the perfume was obtained both by hot and cold enfleurage as well as by distillation. Large quantities of rose water are also made. This consists mainly of the water left after distillation, which still contains some of the essence. Dissolving a small amount of Otto in water sometimes makes it.

Geranium

Pelargonium spp. leaves yield an essential oil after distillation. Geranium oil is widely used as an adulterant of or a substitute for Otto of Roses in making perfumes and soap. *Pelargonium graveolens* is most frequently grown especially in Algeria and Reunion and to a lesser extent in southern France and Spain. Cultural experiments have been made in Florida, Texas and California with *P. odoratissimum,* the rose geranium. The plants are easy to propagate from slips and are productive for 5-6 years after reaching maturity. They must be grown in minimum frost areas. A good grade of oil is obtained from the leaves of this species.

Ylang-ylang

A very valuable oil in the perfume industry, it is added to almost every perfume. The name translates as "flower of flowers." The ylang-ylang tree is an Eastern Asiatic species, *Canaya odorata*. Its yellowish-green, bell shaped flowers have an exceedingly delicate and evanescent fragrance. The oil is also known as Canaga Oil and is derived by simple distillation or extraction from the petals of fully opened blossoms. Most production was originally in The Philippines, but later French colonies in the Indian Ocean dominated the cultivation. This tree grows wild or cultivated in various parts of Southern Asia and the East Indies. The oil first arrived in Europe around 1864 and since that time it has been in great demand despite its high cost.

Cassie or Acacia

Flowers of the sweet acacia, *Acacia farnesiana*, yield an essential oil that is almost as valuable as Ylang-Ylang or Otto of Roses. It is a thorny small tree of the West Indies, but has spread to many tropical and subtropical areas. It is extensively cultivated in southern France, Algeria, Egypt, Syria and India as a source of perfume. The oil is removed from the petals by maceration with cocoa butter or coconut oil, or by extraction. It is similar to the odor of violets and is widely used for sachets, powders and pomades.

Neroli

This oil, obtained from orange blossoms, is extensively used in blends and for mixing with synthetic perfumes. True oil of neroli, or Neroli Bigarade, is distilled from flowers of the bitter orange, *Citrus aurantium*. Neroli Portugal is from the sweet orange, *Citrus sinensis*. Leading production has been in southern France, surrounding Mediterranean areas and in the West Indies, particularly in Haiti.

Other essential oils are also derived from Oranges and used in making perfumes. The leaves and twigs and sometimes immature fruits, supply Petitgrain Oil. This adds a pleasant bouquet to scents, cosmetics and soap. Paraguay has been the main producer. Bitter or sweet oranges are used and the oil is extracted by distillation. Oil of Orange is obtained by expressing the ripe orange peel, but it results in an inferior grade.

Bergamot

This is greenish oil that is expressed from the rind of the bergamot (*Citrus aurantium subsp. bergamia*). It has a soft sweet odor and has been widely used in the United States for adding scent to toilet soaps and in mixed perfumes. Italy and Sicily have been the chief exporters.

Orris

Rhizomes of *Iris pallida*, *I. florentina* and allied species contain an essential oil that has the odor of violets. Tincture of Orrisroot has been used to adulterate pure extract of violets and the powdered root is the basis of violet powder. Cultivation is in Southern Europe, Iran and northern India. Italian orrisroot is superior to other sources. The rhizomes are peeled and dried in the sun, and the odor gradually develops. Orris is also used as a flavouring substance.

Calamus

The roots of Calamus are the sweet and aromatic rhizomes of the sweet flag, *Acorus calamus*. It is a common plant of marshy ground in Europe, Asia and America. In powdered form Calamus is used for sachet and toilet powders, while the distilled oil is used in making perfumes. It has also been used for medicinal and flavouring purposes. The candied root was once a popular confection.

Grass Oils

Several important essential oils are derived from grasses and used in the perfume industry. The genus *Cymbopogon* (formerly *Andropogon*) is especially rich in perfume species.

Oil of Citronella

The oil is distilled from the leaves of *Cymbopogon nardus*. Java and Sri-Lanka have produced most of the world supply. The pale-yellow oil is inexpensive and used for making soaps and perfumes and as an insect repellent. The oil contains 80-90 percent geraniol, and is therefore an important substitute for Otto of Roses. Citronella was also introduced into Central America where a considerable industry developed in Guatemala and Honduras with over 4,500 acres under cultivation by 1952. The crop is harvested by hand. Cutting stimulates growth and a new crop is available in three months. The oil is extracted by steam distillation.

Lemon-grass Oil

Leaves of *Cymbopogon citratus* yield reddish-yellow oil with a strong odor and taste of lemons upon distillation. There is a very high content of citral in the leaves (70-80%). It is used in soaps and medicine. Citral is extensively used in perfumes, bath salts, cosmetics and toilet soaps and as a food flavouring. It is also the source of the aromatic substances known as ionones, which have many uses. One of the ionones is required in the synthesis of Vitamin A; another is the raw material for synthetic violet. Lemon grass is common in the eastern tropics and is cultivated in Sri-Lanka, East Africa, India, the Congo and Madagascar. It was successfully introduced in the Western Hemisphere where large quantities of the oil were exported from Haiti, Brazil, Guatemala, El Salvador and other Neotropical areas.

Palmarosa / Ginger-grass Oils

These are nearly identical oils that have been used as adulterants of Otto of Roses. They contain a high amount of Geraniol. *Cymbopogon martinii* and related species are cultivated in India as a source and have been exported in large quantities.

Oil of Vetiver

The roots and rhizomes of the khuskhus plant, *Vetiveria zizanioides*, supply this oil. It is a native of Bengal and India, but has been grown throughout the tropics and subtropics. The roots have a very sweet scent and are made into mats, fans, screens, awnings, sunshades, baskets, sachet bags and pillows. The leaves are odourless. The plant was introduced into the West Indies and Louisiana and is widely used as an ornamental plant. It has escaped cultivation and become naturalised in many areas. On distillation the roots yield oil that is similar to citronella and which has been used for making expensive perfumes, soaps and in medicine. It is one of the best fixatives.

Oil of Bay

Leaves of *Pimenta racemosa* yield oil on distillation, which is used in perfumery and in the preparation of bay rum. The plant is native to the West Indies where the industry is located. Originally the leaves were distilled in rum and water, but now the oil is dissolved in alcohol and various aromatic substances are mixed in. Bay rum has soothing and antiseptic properties.

Lavender

Lavender perfumes are very old and were used by the Romans in their baths. It is still one of the most important scents. The true lavender plant, *Lavandula officinalis*, is native to Southern Europe, where it occurs on dry, barren soil. It is a low shrub with terminal spikes of very fragrant bluish flowers. There are many horticultural forms and hybrids occur. Lavender is grown in southern France at altitudes of 1500-1800 ft. Large amounts are also raised in England. Lavender has a clean odor and the dried flowers are used in sachets and for scenting chests and drawers. The oil is important in the manufacture of Eau de Cologne and other perfumes and is also used in soaps, cosmetics and medicine as a mild stimulant. Lavender water, a mixture of the oil in water and alcohol, is popular in England (Yardley brand).

Spike Lavender

This lavender, *Lavandula latifolia*, is coarser and yields an inferior grade of oil. It can be grown at lower altitudes than true lavender and is extensively cultivated in France and Spain. It is used in perfumes and cosmetics and to flavour meat sauces known as aspic.

Violet

One of the most popular perfumes is made from violets. Blue and purely double varieties of *Viola odorata*, native to Europe, are grown mainly in the vicinity of Nice. Solvents or maceration with hot fats extracts the oil. It occurs in such minute amounts that 15 tons of flowers are required to obtain only one pound of oil. Genuine violet perfume is rare and expensive, and it has been almost entirely replaced by synthetic products derived from ionone.

Jasmine

A highly esteemed perfume, jasmine is cultivated in southern France and surrounding areas. The main source is *Jasminum officinarum* var. *grandiflorum*, which is usually grafted on a less desirable variety. The flowers are picked as soon as they are open and the oil is extracted by enfleurage.

Carnation

There are thousands of horticultural varieties of carnation all derived from *Dianthus caryophyllus*, a species of southern Europe, Northern Africa and tropical Asia. The most conspicuous blossoms and colour give the least odor. Therefore, for the perfume industry varieties with less conspicuous blossoms are used. Solvents extract the oil. Synthetic carnation oil predominates in world markets.

Rosemary

Rosemary, *Rosmarinus officinalis*, is a native of the Mediterranean region. It has long been a favoured sweet-scented plant and has been important in the folklore of many countries. It is one of the least expensive and most refreshing odours. The plant is a small evergreen shrub that is cultivated in Europe and the United States. The oil is extracted by distillation of the leaves and fresh flowering tops or by extraction. It is used in Eau de Cologne, toilet soap and medicine. The leaves are valuable as a spice.

Hyacinth

Hyacinthus orientalis is native to Western Asia and Asia Minor. It was introduced into Europe in the 16th Century where it was grown as an ornamental plant especially in The Netherlands. It is also a familiar species in North America. Hyacinths are grown for perfume in southern France. The odor is heavy, sweet and quite overpowering. Solvents are used to obtain the oil, which is generally greatly diluted.

Oak Moss

Oak Moss, also called Mousse de Chene, is a valuable addition to the raw materials of the perfume industry. It comprises various lichens that grow on the bark of trees. The main sources are European species of *Ramelina* and *Evernia*, particularly *R. calicaris, E. furfuracea* and *E. prunastri*. These lichens contain oleoresins that are extracted by means of solvents. After they have been collected, the lichens are dried. Then the perfume develops in storage. Oak moss not only has a heavy, penetrating odor and blends well, but also has a high fixative value. It is an essential element in lavender perfumes and soap and in the better grades of cosmetics.

Linaloe or Bois de Rose

Several sources of this very aromatic substance occur. Mexican linloe is distilled from wood chips of two species, *Bursera penicillata* and B. glabrifolia. Cayenne linaloe or Bois de Rose is derived from *Aniba panurensis* of the Guianas, while Brazilian Bois de Rose is from *A. rosacodora var. amazonia*, a tree in the lower Amazon basin. The product is widely used in perfumes, soaps, and cosmetics and for flavouring of foods and beverages.

Sandalwood

The oil is obtained by distillation from the wood of *Santalum album* and related species. The tree grows wild in India and other parts of Southeastern Asia and is cultivated in many other areas. The oil is used throughout the Orient as a perfume and also in medicine. It is an excellent fixative and is used in blends. The sweet-scented wood is made into chests and boxes. Demand for sandalwood has been very great, resulting in the eradication of the species in many areas. Several substitutes have been used.

Patchouli

This oil is obtained from the flesh leaves and young buds of Pogostemon cablin. The plant is a small shrub that grows wild in Southeastern Asia and is cultivated in China. The leaves are fermented in piles and are then distilled. The dark-brown oil has a powerful odor, resembling that of sandalwood. It is one of the best fixatives for heavy perfumes. It is also used in soaps, tobacco and hair tonics. It imparts the characteristic odor to cashmere shawls, which are shipped in patchouli-scented containers.

Champaca

This oil constitutes one of the most famous perfumes of India and other oriental countries. It is obtained from *Michelia champaca*, a large tree of the eastern tropics. The conspicuous yellow flowers are very fragrant and are frequently worn by the natives. The oil is removed from the flowers by maceration or extraction and rivals ylang-ylang in its delicious fragrance.

Misc Perfume Sources

Some of the gum resins, mainly frankincense and myrrh, have been used in perfumery for thousands of years. Other garden flowers that are cultivated for their perfume include Heliotrope (*Heliotropium arborescensj*), Lily of the Valley (*Convallaria majalis*), Jonquil (*Narcissus jonquilla*), Mignonette (*Reseda odorata*), Narcissus (*Narcissus tazetta*), Clary Sage (*Salvia sclarea*) and Tuberrose (*Polianthes tuberosa*). Oils from caraway, anise, cassia, cinnamon, clove, peppermint, lemon, thyme wintergreen and zedoary have also been used in the perfume industry, but these are discussed under other categories.

Camphor

Camphor is an important essential that is used in industry. Commercial camphor, called camphor gum, consists of tough, white translucent masses or granules with a penetrating odor and pungent aromatic taste. It is solid at room temperature, thus bearing the same relation to the other essential oils that vegetable fats do to the fatty oils. It volatilises very slowly.

The oil is obtained by distillation of the wood of the camphor tree, *Cinnamomum camphora*, and native to China, Taiwan and Japan. This is a very tall and striking tree with shiny, dark evergreen leaves. The tree has been widely introduced into tropical and subtropical regions, mainly as an ornamental plant. The camphor industry is centred in Taiwan. Earlier crude methods of obtaining camphor were very destructive and the existence of the tree was threatened. Finally only trees 50 years of age or older were used and every stages in the process was carefully regulated. The wood is reduced to chips or ground to a fine powder and the leaves are also ground up. This is then distilled with steam for several hours and the crude camphor crystallises on the walls of the still. This is removed and must be purified before it is ready for market. Synthetic camphor from pinene, a turpentine derivative, gradually dominated the market.

The principal use of camphor has been in the manufacture of celluloid and various nitrocellulose compounds. It also has a wide range of medicinal uses, both internally and externally. It is also used in perfumery. Borneo camphor, obtained from *Dryobalanops aromatica* of the east Indies, has been used as a substitute.

Cedarwood Oil

Cedarwood oil, along with clove and bergamot oils, is one of several of the essential oils have a high refractive index and are valuable as clearing agents in the preparation of permanent microscopic mounts and for use with oill-immersion lenses. This inexpensive oil is obtained by steam distillation from the heartwood of the eastern red cedar, *Juniperus virginiana*, and related species. Wood chips, sawdust, waste from the lead pencil and other industries, old stumps, roots and even fence rails have been utilised. Cedarwood oil is also used in perfumery, soaps, deodorants, liniments, cleaning and polishing preparations and as an adulterant of expensive sandalwood and geranium oils. It has insecticidal properties and is used as a moth repellent and in fly sprays.

Misc. Oils

Essential oils are also useful as solvents for paints and varnishes. The most important of these is oil of turpentine. Various other oils, mainly eucalyptus oil from *Eucalyptus dives*, are employed in the flotation process for the separation of minerals from their ores. Still other volatile oils have been used in the preparation of cleaning materials and other industrial purposes.

The Absorption and Effects of Essential Oils

Glandular

Essential oils probably exert their most powerful and direct pharmacological effects systemically via the blood supply to the brain. They also have an indirect effect via the olfactory nerve pathways into the brain. Essential oil fragrances are absorbed through blood circulation and nerve pathways from the sinuses into the central glands of the brain, which control emotional, neurological, and immunological functions.

Skin

Essential oils are absorbed in minute quantities through the skin, depending on the oil, dilution, and application (carrier oil, compress, etc). Many of the indications for specific oils include various skin conditions.

Respiratory

Essential oils are inhaled during treatment, which have a direct effect of the sinuses, throat, and lungs. Many essential oils are specific medicines for respiratory conditions.

Circulation

Many essential oils have beneficial effects on circulatory problems, both through dermal and respiratory absorption. These oils enhance the circulation stimulating effects of massage.

Adulteration and Contamination

The 10 Most Important Points To Know Before You Purchase Essential Oils:

1. "Pure": In the US, the term "pure" has no legal meaning and is often applied to just about anything.
2. Synthetic Fragrances: Certain oils do not exist in a natural state, and are only available as synthetic fragrances or "bouqueted" fragrances (combination of essential oils, absolutes, and synthetics). These include honeysuckle, linden, gardenia, frangipani.
3. Adulteration: The more expensive an oil, the more risk of adulteration. Some oils are highly adulterated, such as melissa (lemon balm), rose, and sandalwood.
4. Chain of Supply: The fragrance industry has many levels of buyers and suppliers. The more levels that are involved, the more there is risk of adulteration. Large volumes of oils are sold as "genuine" and "pure," which are not. False advertising is rampant in the aromatherapy world. It is best to get oils directly from the distiller.

   Pesticides: Some pesticides are carried over in the extracting process, some are not.

   Expressed citrus oils contain pesticide residues.

5. Grades: Lower grades of oils are frequently sold as higher. A good example is ylang ylang.
6. Extenders: Many oils are "extended" using synthetic or natural solvents. Expensive oils are frequently extended with jojoba. Some oils are extended to make them more pourable, like benzoin; the solvent is frequently questionable.
7. Bulking: Bulking is the post-distillation combining of oils from one or more species, or loading plants of the same species from different harvests into the still together. Dried plant material from different years may be bulked with fresh.

   Bulking is done to make the product cheaper and/or to make it conform to some standard desirable to the fragrance or flavouring industries.
8. Rectified or Redistilled: Oils that have had natural components removed from them: terpene-less oils, furocoumarin-free oils.
9. Folded: Oils, (usually citrus) that have been redistilled a number of times to remove more of the monoterpenes (usually) to make the oil more desirable for the flavouring industry.
10. Reconstituted: Oils that have had natural or synthetic chemical components added to them after distillation.

The Safe Use of Essential Oils

In general, when used properly essential oils are quite safe and highly beneficial. However, because their uses are still relatively unknown, people can and do hurt themselves by using these highly concentrated botanical substances improperly.

Toxicology and Safety

- Do not use essential oils internally.
- Do not apply directly to skin; always dilute with carrier oil.
- Keep out of reach of children.
- Avoid contact with eyes and mucous membranes.
- Do not use citrus oils before exposure to UV light.
- Use only pure essential oils; avoid synthetic fragrances.
- Do not use essential oils on infants, children, pregnant women, the elderly, or those with serious health problems, without advanced medical study.
- Avoid prolonged exposure without ventilation.
- Store essential oils and carrier oils properly to avoid degradation and rancidity.

Do not use essential oils internally.

There are two exceptions to this rule.

- The first is properly administered dosages of essential oil medications prescribed by a licensed physician. This is now occurring in certain European clinics, but is rarely available in the US. People should avoid using essential oils internally if prescribed by a lay practitioner, especially if the practitioner's education is primarily from the marketing perspective rather than clinical.
- The second exception is biocompatible levels of essential oil ingestion when taken as part of the diet. A good example of this is oregano oil. Oregano oil is widely marketed for internal consumption, with numerous claims made about its therapeutic efficacy. In actual practice, the internal consumption of this oil frequently causes the typical symptoms associated with the ingestion of essential oils, such as extreme gastric hyperacidity. On the other hand, the use of oregano as a fresh herb, steamed at the end of food preparation, provides all the benefits of oregano oil at a biocompatible level, with none of the gastric dangers.

Should accidental ingestion of any significant amount of an essential oil occur, immediately call your local Poison Control Centre. Do not induce vomiting. Do not give water if breathing or swallowing is difficult. Do not apply directly to skin; always dilute with carrier oil. Essential oils are very concentrated. Dilute all essential oils before applying to the skin, either in a fatty oil, or in water as when used on a compress.

There are two exceptions to this rule. The first is the use of attars as perfumes. Because the floral essences are distilled into a base of sandalwood oil, the sandalwood oil acts as a carrier which dilutes the potency of the pure essential oil. The second is the reasonable use of mild essential oils that have a well documented history of safety. The best example of this is lavender; however, even lavender can be problematic for some people. Skin reactivity is becoming more of a problem as synthetic aroma chemicals become more common adulterants in the essential oil industry.

A general rule is to never apply more than one to two drops of undiluted oil to the skin. Patch testing is always advisable. For people with sensitive skin, always test a small area with a diluted oil before applying over a larger area. For general non-medical use, avoid essential oils with highly sensitive skin and with any instances of skin allergies, severe inflammation and dermatitis. Pure essential oils are much less dangerous than synthetic aroma chemicals.

Skin Reactions

Skin reactions are dependent on the type of oil, the concentration of the oil, and the condition of the skin. It is best to check with clients to determine any prior history of skin reactions before using oils, either for dermal or respiratory applications.

Old and oxidised oils are more prone to cause reactions, especially rashes. Refrigerate fatty carrier oils to prevent rancidity. Essential oils generally have a shelf life of one to three years. Some get better with age, such as sandalwood, vetiver, and patchouli. The citrus oils are most prone to degradation, and should be used within one year.

Skin reactions to essential oils can take three forms:

- Irritation: A small number of oils are strongly or severely irritant. These include horseradish, mustard, garlic, and onion (which are rarely used in aromatherapy practice). Some oils used in massage practice can be moderately irritant, such as cinnamon, clove, fennel, and verbena. These oils should be used cautiously or avoided in cases of skin sensitivity.
- Sensitisation: Skin sensitisation means an allergic skin reaction; this usually manifests as a rash. There are relatively few oils used in a typical massage practice that will produce sensitisation under normal applications in a carrier oil. However, there are a number of reports on Pubmed of allergic reactions to essential oils. These include contact dermatitis, eczema, asthma, and pruritic erythematous eruptions. These cases were predominantly among those who used essential oils professionally for long periods of time, such as massage therapists and estheticians. The cases frequently involved exposure to numerous essential oils, and it is also likely that the quality of the oils was poor.
- Phototoxicity: Some essential oils can strongly increase sensitivity to sunlight when applied to the skin. This is especially dangerous when applied undiluted to the skin, but even low concentrations in a carrier oil can cause problems if followed by exposure to sun or tanning lamps. Phototoxicity will be much stronger directly after application of the oil, and will gradually decrease over an eight to twelve hour period; if higher than normal concentrations are used it can be longer.

Most of the phototoxic oils are also photocarcinogenic. The most common oils which cause phototoxicity are the citruses; bergamot is the most reactive. Some citruses are phototoxic if expressed, but not if distilled, such as lemon and lime. Other oils include marigold (tagetes), verbena, and angelica. The best practice is to use proper dilutions, avoid direct exposure to UV rays after application, and avoid the use of citrus oils if exposure will be occurring after treatment.

The best treatment for skin irritation from essential oils is to apply a fatty oil, such as coconut, which will dilute the impact of the essential oils.

Avoid Contact with Eyes and Mucous Membranes

If an essential oil gets into the eye, do not rub it. Saturate a cotton ball with milk or vegetable oil and wipe over the area affected. In severe instances flood the eye area with lukewarm water for fifteen minutes. Take special precautions with applications near delicate skin areas. Use only pure essential oils; avoid synthetic fragrances.

Avoid Prolonged Exposure without Ventilation

Overexposure to essential oils, especially in confined areas, can cause dizziness, nausea, light headedness, headache, blood sugar imbalances, irritability, euphoria. When exposed to high levels of essential oils make sure to keep the room well ventilated.

Store essential oils and carrier oils properly to avoid degradation and rancidity. Air, heat and light degrade essential oils. Store essential oils in a cool, dark room and always keep your oils tightly sealed. Do not use essential oils on infants, children, pregnant women, the elderly, or those with serious health problems, without advanced medical study.

General Advice

Before experimenting with an oil, become familiar with it's properties, dose, and precautions. When in doubt about a condition or an oil, consult a qualified medical specialist. If in doubt use safe, non-irritating essential oils and dilute them with a carrier oil before using.

Peppermint

Peppermint (*Mentha piperita*), a popular flavouring for gum, toothpaste, and tea, is also used to soothe an upset stomach or to aid digestion. Because it has a calming and numbing effect, it has been used to treat headaches, skin irritations, anxiety associated with depression, nausea, diarrhea, menstrual cramps, and flatulence. It is also an ingredient in chest rubs, used to treat symptoms of the common cold. In test tubes, peppermint kills some types of bacteria, fungus, and viruses, suggesting it may have antibacterial, antifungal, and antiviral properties. Several studies support the use of peppermint for indigestion and irritable bowel syndrome.

Indigestion

Peppermint calms the muscles of the stomach and improves the flow of bile, which the body uses to digest fats. As a result, food passes through the stomach more quickly. However, if your symptoms of indigestion are related to a condition called gastroesophageal reflux disease or GERD, you should not use peppermint.

Flatulence/Bloating

Peppermint relaxes the muscles that allow painful digestive gas to pass.

Irritable Bowel Syndrome (IBS)

Several studies have shown that enteric coated peppermint capsules can help treat symptoms of IBS, such as pain, bloating, gas, and diarrhea. (Enteric coated capsules keep peppermint oil from being released in the stomach, which can cause heartburn and indigestion.) However, a few studies have shown no effect. One study examined 57 people with IBS who received either enteric coated peppermint capsules or placebo twice a day for 4 weeks. Of the people who took peppermint, 75% had a significant reduction of IBS symptoms. Another study comparing enteric coated peppermint oil capsules to placebo in children with IBS found that after 2 weeks, 75% of those treated had reduced symptoms. Finally, a more recent study conducted in Taiwan found that patients who took an enteric coated peppermint oil formulation 3 - 4 times daily for one month had less abdominal distention, stool frequency, and flatulence than those who took a placebo. Nearly 80% of the patients who took peppermint also had alleviation of abdominal pain.

Itching and Skin Irritations

Peppermint, when applied topically, has a soothing and cooling effect on skin irritations caused by hives, poison ivy, or poison oak.

Tension Headache

One small study suggested that peppermint applied to the forehead and temples helped reduce headache symptoms.

Colds and Flu

Peppermint and its main active agent, menthol, are effective decongestants. Because menthol thins mucus, it is also a good expectorant, meaning that it helps loosen phlegm and breaks up coughs. It is soothing and calming for sore throats (pharyngitis) and dry coughs as well.

Plant Description

Peppermint plants grow to about 2 - 3 feet tall. They bloom from July through August, sprouting tiny purple flowers in whorls and terminal spikes. Dark green, fragrant leaves grow opposite white flowers. Peppermint is native to Europe and Asia, is naturalised to North America, and grows wild in moist, temperate areas. Some varieties are indigenous to South Africa, South America, and Australia. The leaves and stems, which contain menthol (a volatile oil), are used medicinally, as a flavouring in food, and in cosmetics (for fragrance).

Precautions

The use of herbs is a time honoured approach to strengthening the body and treating disease. Herbs, however, can trigger side effects and interact with other herbs, supplements, or medications. For these reasons, you should take herbs with care, under the supervision of a health care provider.

Do not take peppermint or drink peppermint tea if you have gastroesophageal reflux disease (GERD — a condition where stomach acids back up into the esophagus) or hiatal hernia. Peppermint can relax the sphincter between the stomach and esophagus, allowing stomach acids to flow back into the esophagus. (The sphincter is the muscle that separates the esophagus from the stomach.) By relaxing the sphincter, peppermint may actually make the symptoms of heartburn and indigestion worse.

Peppermint, in amounts normally found in food, is likely to be safe during pregnancy, but not enough is known about the effects of larger supplemental amounts. Speak with your health care provider. Never apply peppermint oil to the face of an infant or small child, as it may cause spasms that inhibit breathing. Peppermint may make gallstones worse. Large doses of peppermint oil can be toxic. Pure menthol is poisonous and should never be taken internally. It is important not to confuse oil and tincture preparations.

Menthol or peppermint oil applied to the skin can cause a rash.

Possible Interactions

Cyclosporine : This drug, which is usually taken to prevent rejection of a transplanted organ, suppresses the immune system. Peppermint oil may slow down the rate at which the body breaks down cyclosporine, meaning more of it stays in your bloodstream. Do not take peppermint oil if you take cyclosporine.

Drugs that Reduce Stomach Acid : If peppermint capsules are taken at the same time as drugs that lower the amount of stomach acid, the enteric-coated peppermint capsules may dissolve in the stomach instead of the intestines. This could mean the effects of peppermint are lessened. Take peppermint at least 2 hours before or after an acid-reducing drug. Antacids include:

- Famotidine (Pepcid)
- Cimetidine (Tagamet)
- Ranitidine (Zantac)
- Esomeprazole (Nexium)
- Lansoprazole (Prevacid)
- Omeprazole (Prilosec).

Drugs that Treat Diabetes : Test tube studies suggest peppermint may lower blood sugar, raising the risk of hypoglycemia (low blood sugar).

Medications Changed by the Liver : Since peppermint works on the liver, it may affect medications that are metabolised by the liver (of which there are many). Speak with your health care provider.

Antihypertensive Drugs (Blood Pressure Medications) : Some animal studies suggest that peppermint may lower blood pressure. If you take medications to lower blood pressure, taking peppermint also might make their effect stronger.

Essential oils are concentrated liquids that are derived from plants. They are a type of hydrophobic liquid, which, as you may have guessed, means that they repel water. They contain volatile aromatic compounds. For that reason, essential oils are also called volatile oils or ethereal oils. Essential oils shouldn't be confused with fatty acid oils such as soya, rape seed or evening primrose; their chemical properties and biochemical origins are quite different. Sometimes essential oils are called "oil of", as in "oil of clove." The word "essential" in this case means that the oil is the fragrant essence of the plant.

Here are some sample plant parts and what essential oils are derived from them:

- Flowers - Chamomile. Clary sage, Clove, Geranium, Hyssop, Jasmine, Lavender, Manuka, Marjoram, Orange, Rose, Ylang-ylang
- Leaves - Basil, Bay leaf, Cinnamon, Common sage, Eucalyptus, Lemongrass, Oregano, Patchouli, Peppermint, Pine, Rosemary, Spearmint, Tea tree, Thyme,
- Berries - Allspice, Juniper
- Seeds - Almond, Anise, Celery, Cumin, Nutmeg oil
- Peel - Bergamot, Grapefruit, Lemon, Lime, Orange, Tangerine
- Resin - Frankincense, Myrrh
- Bark - Cassia, Cinnamon, Sassafras
- Wood - Camphor, Cedar, Rosewood, Sandalwood
- Root - Valerian
- Rhizome – Ginger.

As you may have noticed, some essential oils, such as cinnamon, may come from a combination of parts, such as leaves and bark. The oils sources in plants are extracted from cavities or ducts plus glands or hairs. Clusters of cells just below the plant's epidermis create cavities and ducts. Glands can be found on lavender florets; and geranium and mint feature leaf hairs. Essential oils have been used since antiquity, for everything from perfume to flavouring agents to disease remedies—and even for embalming mummies! Medical claims are now regulated by most countries. Essential oils are used by many industries in today's world, including paint, manufacturing, food and beverage production, pharmaceuticals, toiletries and hygiene products. The popularity of essential oils has surged amongst consumers in the past few decades with the practice of aromatherapy.

Essential Oils Production Process

Essential oils are produced by distillation, expression or solvent extraction. The most common essential oils, such as lavender, peppermint or eucalyptus, are produced using distillation. Distillation involves placing raw plant material into an alembic, which is placed over water. An alembic is a bulbous container with a tube descending from its top that's attached to a smaller bulbous container. The raw plant material can consist of flowers, leaves, roots, seeds, wood or bark.

The alembic is placed over water, and the water is heated. As the water heats, steam passes through the raw plant material; this vapourises the volatile plant compounds. The vapours then flow through a coil where they condense into liquid form; the liquid is collected in the smaller receiving vessel of the alembic. Of course, producers of therapeutic-grade essential oils utilise distillation chambers that are much larger than alembics! For example, Young Living essential oils, one of the largest essential oil producers in the world, uses numerous distillation vats with an overall boiler capacity of more than 115,000 litres.

Most essential oils are distilled in one single process, although there are some exceptions. For example, ylang-ylang is produced through a special 22-hour distillation process called fractional distillation. The water that is recondensed during the distillation process is called a hydrosol, hydrolat, herbal distillate or plant water essence. These are often sold as different fragrant products; you've probably seen rose water, lavender water and orange blossom water. Rose water is used extensively in India in cooling yogurt drinks called Lassi, plus in religious rituals. But believe it or not, some hydrosols possess unpleasant odours! Of course, these are not sold.

Some essential oils, namely those derived from citrus peels, are expressed mechanically. This is called cold-pressed—just like olive oil! Prior to the invention of the alembic and distillation in 800 AD by the Arab alchemist Jabir ibn Hayyan, *all* essential oils were produced via pressing.

Some plants, particularly their flowers, contain very little volatile oil, so the expression process is ineffective. And some plants' chemical components are too delicate for the high heat required for the distillation process. That's where solvent extraction comes in. Solvents, such as supercritical carbon dioxide or hexane, are used to extract the volatile oils. Extracts from hexane are called concretes; these are a combination of essential oil, oil-soluble plant material, resins and wax. Although concretes are quite fragrant, they also contain large amounts of non-fragrant resin and wax. Ethyl alcohol dissolves the fragrant low-molecular weight compounds to extract fragrant oil from concretes.

Then the ethyl alcohol is removed during a second distillation, which leaves what is called the absolute. Using supercritical carbon dioxide helps prevent petrochemical residues from appearing in the final product. Supercritical carbon dioxide doesn't directly yield an absolute; rather, it extracts the wax and essential oils that comprise the concrete. Then liquid carbon dioxide is used in conjunction with lowering the extraction temperature, which separates wax from the essential oils. The lower temperature prevents chemical compounds from decomposing to yield a superior product. Once the extraction process is complete, the pressure is reduced so that the carbon dioxide reverts to gas form, which leaves no residue.

Essential Oils Safety Precautions

Generally speaking, full-strength essential oils should not be applied directly to the skin because they're so concentrated. This is called their "neat" form. Certain oils may instigate severe irritation or an allergic reaction. Rather, essential oils should be blended with a vegetable carrier oil before application. These carrier oils are called base oils or fixed oils. Preferred carrier oils are organic and/or cold pressed.

Commonly used carrier oils include almond, apricot kernel, avocado, grapeseed, jojoba, hazelnut, sesame or wheat germ. A typical ratio of essential oil to carrier oil usually ranges from 0.5 to 5 percent, depending on the use; a 3 to 5 percent dilution is 3 drops of essential oil per 1 teaspoon of carrier oil. As a matter of fact, industrial handlers of essential oils consult material safety data sheets about the potential hazards and handling requirements for certain essential oils. Also, some essential oils, such as those derived from citrus peels, are photosensitisers; in other words, they increase the likelihood of becoming sunburned.

Essential oils are remarkably potent. So here are some important safety precautions:

- Never assume that an essential oil possesses the same properties as its plant
- If you have sensitive skin, epilepsy, heart ailments, kidney problems, or any serious medical condition, do not use essential oils except under the supervision of your physician
- Read and follow all label directions and cautions
- Do not apply neat, or undiluted, essential oils directly to the skin; they must be diluted with a carrier oil
- Always conduct a test of diluted essential oil on the inner arm before using; do not use if redness or irritation occurs
- Keep essential oils out of eyes, ears, nose, mouth, wounds and any body opening
- Keep out of reach of children and pets; do not use on children or pets except under the supervision of a pediatrician or veterinarian
- Keep away from sources of flame, heat and ignition
- You should never use these essential oils, as they possess very high oral and dermal toxicity: bitter almond, buchu, camphor, sassafras, calamus, horseradish, mugwort, mustard, pennyroyal, rue, savin, savory, southernwood, tansy, wormseed and wormwood

- These essential oils are potential skin irritants: allspice, basil, bay, birch, bitter almond, black pepper, cassia, cinnamon leaf, cinnamon bark, citronella, clove bud, costus, cumin, fennel, fir needle, ginger, lemon, lemongrass, lemon verbena, melissa, myrrh, oak moss, orange peppermint, oregano, parsley seed, peppermint, pimento berry, pine, tagetes, tea tree, red and wile thyme and wintergreen
- You should never use these essential oils on the skin: ajowan, cassia, elecampane and origanum
- These essential oils will render skin more sensitive to sunlight: angelica, bergamot and all citrus oils
- You should not use these essential oils during pregnancy: aniseed, basil, bitter almond, birch, cedarwood, clary-sage, cypress, basil, marjoram, clove, fennel, hyssop, jasmine, juniper, lemon, myrrh, lemon verbena, peppermint, rose, rosemary, sage, sweet marjoram, thyme and wintergreen
- You should not use these essential oils if you have high blood pressure: hyssop, rosemary, sage or thyme
- You should not use these essential oils if you have epilepsy: Sweet fennel, hyssop, sage, rosemary and wormwood

There are a few essential oils that can be used for dietary purposes. However, you should never ingest an essential oil without the recommendation of a licensed professional, especially during pregnancy, as they can cause miscarriage. That's because some essential oils are internally toxic, such as eucalyptus. Remember, essential oils are a very highly concentrated form of the plant; their properties can change.

Taruna Oils

Taruna Oils offers therapeutic grade aromatherapy & essential oils for mind, body & spirit. Freshly cultivated and steam distilled, the essential oils in Taruna Oil's collection are life-enhancing gifts from nature.

3

Medicinal Oil Plants

The use of plant-based products for disease prevention and treatment has become increasingly popular in many societies.

The World Health Organization (WHO) has estimated that about 80% of the population in developing countries rely chiefly on traditional medicine for their health care needs, of which a major portion involves the use of plant extracts (Azizol and Jamaludin, 1995). With growing interest in medicinal plants as a source of new pharmaceutical products and the increasing demand for herbal products in Malaysia, it is expected that the demand for raw materials will also increased.

Since most of the medicinal plants resources are from natural resources and many plant species are now facing extinction, it is of necessary to domesticate and cultivate selected species from both forest and non-forest areas. The success of such domestication programme will assist in the conversation of plant genetic resources, avoid further depletion and meet the demand for raw materials from the herbal industries.

Medicinal and Aromatic Plants of Malaysia

In Malaysia, the biodiversity of plant resources offers some 12 500 species of flowering plants and 5000 species of cryptograms.

About 2000 species are recognized for their medical properties and they are still being used among certain communities (Latin, 1994).

Some of these plants that are used as traditional medicines are also used as common spices or food additives. Species which have been commonly used for herbal preparations.

Table: *Some Commonly Used Medicinal Plantsa*

Species	***Local name***	***Common use(s)***
Eurycoma longifolia	Tongkat Ali	Health tonic
Labisia pumila	Kacip Fatimah	Post-partum preparation
Centella asiatica	Pegaga	Health tonic
Curcuma xanthorizza	Temu lawak	Jamu
Andrographis paniculata	Hempedu bumi/ akar cerita	Herbal tea
Zingiber zerumbit	Lempoyang	Jamu
Eugenia aromatica	Cengkih	Toothpaste
Mentha arvensis	Pudina	Toothpaste
Curcuma domistica	Kunyit	Cosmetic, food additive
Cassia alata	Gelenggang	Antiseptic
Smilax myosotiflora	Ubi jaga	Health tonic
Morinda citrifolia	Mengkudu	Health tonic, past-partum preparation
Leptospermum flavescens	China maki	Health tonic
Fibraurea odoratum	Pokok kapal terbang	antiseptic

* Burkill (1996); Perry and Metzer (1980)

Table: *Some commonly used aromatic plantsa*

Species	***Local name***	***Common use(s)***
Piper nigrum	Lada hitam	Flavour
Cympogon nardus	Serai wangi	Cosmetics, insect repellant
Kaempferia galanga	Cekur	Spice
Lawsonia inermis	Inai	Cosmetics
Melaleuca cajuputi	Gelam	Analgesic
Baeckea frutescens	Rempah gunung	Fragrance
Ocimum basilicum	Selasih	Cosmetics
Jasminium sambac	Melati	Fragrance
Michelia champaca	Cempaka	Cosmetics
Blumea balsamifera	Sembung	Health tonic, lotion
Cinnamomum zeylanicum	Kayu manis	Spice, fragrance
Cinnnamomum sintoc	Medang sintoc	cosmetics

* Burkill (1996); Perry and Metzer (1980)

The aromatic plants like pepper, turmeric, ginger, cinnamomum, lemon grass etc. are exclusively used in the house-hold sector as natural food flavouring. Some of the aromatic plants which have the potential to be used in industry. Lately, the demand for natural aromatic resources is increasing in the international essential oil market. Essential oils which are obtained from the bark, leaves, flowers and fruits are natural sources for fragrance, flavour, species and medicine.

Medicinal and Aromatic Plant Industries

Statistics have shown that between 1986 and 1996, the total import value of medicinal and aromatic plants increased from RM 141 million and RM 431 million, respectively. In terms of total export, there was are significant increase from RM 5.9 million to RM 63 million over the same period.

The imports of medicinal and aromatic plantscome mainly China, India and Indonesia while exports are largely to Singapore, Phillipines, Australia and Hong Kong. Under the spice category, garlic is the important import item (RM 119.5 milllion in 1996), whereas the export of ginger declined from RM 2.4 million in 1992 to RM 0.6 million in 1996 (Ng and Azmi, 1997).

In industry, medicinal plants and their parts are used in the form of extracts with high and standardised contents of active constituents for the pharmaceutical and natural products industries. Many Malaysian plants which have been traditionally used to treat certain ailments are now being processed using modern technology for the production of functional foods and tonics. These include Allium sativum (garlic), Centella asiatica (pegaga), Eurycoma longifolia (tongkat ali), Labisia pumila (Kacip Fatimah) and Zingiber officinale (halia).

Aromatic plants and their parts are the sources of essential oils, resin, turpentine, flavours and fragrances which can be used in the preparation of traditional medicines as well as in industry. Some of the important essential oils used in medicine are mint oil (Mentha arvensis), peppermint oil (Mentha piperita), eucalyptus oil (Eucalyptus spp.), citranella oil (Cymbapogon nardus) and cinnamon leaf oil (Cinnamomum zeylanicum). As for the international market, keruing oil (Gurjun balsam) has recently been used as fixative in perfumes by manufacturers in Singapore. Agar wood or gaharu from the karas tree (Aquilaria spp.) is sold as oleoresin infiltrated fragrant wood. It was reported that the highest grade gaharu was valued at US$ 27 400 in Dubai (Ng and Azmi, 1997).

Many of our local plants are also rich in aromatic compounds that can be used commercially as flavour and fragrance agents in beverages, food products, confectionery, toothpaste, cosmetics and medicinal preparations. These plants include kunyit (Curcuma domestica), serai makan (Cymbopogon citratus), serai wangi (Cymbopogon nardus), pandan (Pandanus odorus) and keso (Polygonum minus). Given the tremendous diversity of aromatic plant species available in Malaysia and the continuous demand for flavour and fragrance by industries, the economic potential of commercial application of these species is very promising.

Planting of Medicinal Plant

In Malaysia, medicinal plants are generally collected from the wild, with limited cultivation being carried out. This has lead to serious depletion of certain species and put some in danger of extinction. Interest has grown in the cultivation of medicinal plants for herbal use. However, to ensure satisfactory, returns from planting medicinal plants, plant selection must be focused on species highly demanded by the industry. Planting will depend on land availability but the following planting conditions can be recommended.

i. Planting under forest conditions which include virgin forest, logged-over forest and plantation forest: in forest where its resources have been removed (logged over forest), enrichment planting with selected medicinal plants is suitable and beneficial. In plantation forest, planting can be carried out in conventional forest plantations where selected medicinal plants are planted under forest species such as teak, pine, acasia, sentang (Azadirachta exelsa) and kara (Aquilaria malaccensis).

ii. Integration with agricultural crops: other then planting under forest conditions, medicinal plants can also be integrated with other commercial crops such as rubber and oil palm. Such crops are able to provide shade and artificial forest environment to the medicinal species, and

iii. Under open condition: under this condition, medicinal species that have high tolerance to high light intensities such as serai wangi may be planted.

When the Malaysian government announced its intention to boost the herbal industry, the Perak state government allocated a land area of 250 ha in Sg. Klah, Sungkai, Perak for a project to cultivate traditional crops (herb, spices and ulam). A series of long and short term medicinal and aromatic plants, have been identified for planting.

In another similar project, Lembaga Kemajuan Kelantan Selatan (KESEDAR) was selected to lead a project on the mass production of medicinal plants of commercial importance. A total area of 60ha have been allocated in Gua Musang, Kelantan for this project.

Commonly Used Medicinal Plants

The most common diseases during the middle ages were dysentery, epilepsy, influenza, diphtheria, scurvy, typhoid, smallpox, scabies, impetigo, leprosy, pneumonia, stroke, heart attack, scrofula (chronically swollen lymph nodes, later identified as a form of tuberculosis), St. Vitus' Dance (rheumatic chorea, a temporary disorder of the parts of the brain that control movement and coordination), and St. Anthony's Fire (ergotism, caused by ingesting toxic amounts of alkaloids produced by a fungus that infests rye - symptoms included gangrene with burning pain in the extremities, convulsions, hallucinations, and severe psychosis).

Plants listed in this section are those used most frequently to treat these and other illnesses. Quoted excerpts of treatments from medical texts were translated from Middle English.

Aloe: Greek physician Dioscorides recommended aloe externally for wounds, hemorrhoids, ulcers and hair loss. Pliny prescribed it internally as a laxative.

Angelica (Wild Celery): Angelica leaf necklaces were worn as protection against illness and witchcraft. Angelica was reputed to be the only herb witches never used and its presence in a woman's garden or cupboard was successfully used as a defense against charges of witchcraft. Gilbert Anglicus' Compendium of Medicine gives the following prescription for using angelica as a cough remedy: And if the cough is of sticky thick phlegm, give him a syrup made with horehound, the root of fennel, radish, wild celery and anise.

Anise: Hippocrates recommended anise to help clear mucus from the respiratory system. It was also recommended by other renowned physicians as a breath freshener, digestive aid, a cure for "hicket" (hiccups), headache, asthma, insomnia, nausea, lice and infant colic. Anise was so popular in medieval England as a spice, medicine and perfume that in 1305 Edward I placed a special tax on it to raise money to repair London Bridge.

Apple: Hildegard of Bingen prescribed raw apples as a tonic for healthy people and cooked apples as the first treatment for any sickness.

Balm (Melissa): Lemon balm and bee balm were prescribed for nervousness and anxiety in the form of Melissa water or Eau de

Melisse. Recommended for treatment of insomnia, arthritis, headache, toothache, sores, digestive problems and cramps, balm was considered to be something of a cure-all. Greek physicians recommended applying balm leaves to wounds and added the herb to wine to treat a variety of illnesses. Pliny prescribed it to stop bleeding.

Basil: The ancient physicians disagreed on the merits of basil. Dioscorides and Galen warned that taking basil internally would cause insanity and the spontaneous generation of internal worms. Pliny used it to treat stomach ailments. Hildegard of Bingen used basil in a concoction that included powder from a vulture's beak to treat tumors.

Blackberry: Blackberry was also known as "goutberry" as its most popular use was as a treatment for gout. Leaves and bark were chewed for bleeding gums, leaves were applied to the skin to sooth burns and scalds. Blackberry syrup was recommended for treatment of dysentery.

Buckthorn : Buckthorn became popular around the 13th century and was primarily used to purge the body of "foul humors". Buckthorn bark seeped in water produced a powerful laxative. It was also recommended for jaundice, hemorrhoids, gout and arthritis.

Burdock : In the 14th century burdock leaves were used to treat leprosy. Hildegard of Bingen used it to treat cancerous tumours. Burdock root was also prescribed for fever, ringworm and skin infections.

Caraway : Caraway seeds were recommended for the treatment of indigestion, gas and infant colic. Gilbertus Anglicus prescribed the following decoction for a syrup to treat ailments of the lungs: Take barley water which has been strained, raisins, violets, jujube, seed of melon and gourd, wheat starch, licorice, black plums, fennel root, parsley, wild celery, anise, caraway, and make thereof a syrup. That is to say, seep all these in water until the virtue of them be in the water.

Then strain it and add sugar or honey. And then set it over the fire to steep softly. Then take the white of four eggs and beat them well and add them. And always skim it until it is clear. Then take it down and strain it clean so that no dregs remain therein. Put it in a closed vessel.

Chamomile : Used to treat headaches, kidney, liver and bladder problems and as an aid for digestive upsets.

Cinnamon : Hildegard of Bingen recommended cinnamon to treat colds, flu, cancer and "inner decay and slime."

Coltsfoot : Used for treating coughs, wheezing, shortness of breath, fever and inflammation.

Comfrey : Boiling comfrey root in water produced a thick paste in which cloth was soaked and then wrapped around broken bones. Internally, comfrey was used for treating respiratory and gastrointestinal problems.

Dandelion : Dandelion was prescribed to treat colds, boils, ulcers, dental problems, itching, jaundice and gallstones.

Dill : Digestive aid and gas remedy. Also a cure for hiccups. From Gilbertus Anglicus' Compendium of Medicine this treatment for squinacy (quinsy):

But if a postem be of phlegm, after his blood-letting and his purging, make him a gargle of sap from a hazelnut tree, dill, poppy, either the water that bark from a nut or mulberry tree has been seeped in, or the juice of bittersweet, with dried honeysuckle leaves and aloe.

Elecampane : Prescribed to treat coughs, bronchitis, asthma and indigestion.

Fennel : The ancient physicians prescribed fennel to treat infant colic, as an appetite suppressant and recommended the seeds to nursing mothers to boost milk production. Pliny believed that fennel was a cure for eye problems, including blindness. Hildegard of Bingen recommended fennel for treating colds, heart ailments and to aid in good digestion and body odour. Folk healers mixed fennel with strong laxatives such as buckthorn to counteract intestinal cramps.

Fenugreek : Fenugreek seeds mixed with water was used as a salve to soothe inflamed or irritated tissue. Internally, it was used to treat fever and digestion and respiratory ailments. Gilbertus Anglicus considered a plaster using fenugreek (femigreke) in combination with a gargle made of other ingredients to be somewhat of a cure-all:

Good for every postem both within a man's body and without: Take the root of hollyhock and lily roots and seep them in water. Then crush them with fresh grease and butter and add meal of flax seed (linseed) and fenugreek and snails and crush them together. And give him a gargle of vinegar that barley has lain in and water that pomegranate or sumac or roses or oak galls or lentils have soaked in.

Garlic : Greek and Roman physicians recommended garlic for infections, wounds, cancer, leprosy, heart problems, colds, and epilepsy among many other ailments. In the middle ages the upper class shunned the use of garlic, but the peasantry viewed it as a preventative medicine and cure-all.

Horehound : First used in ancient Rome as an ingredient in poison antidotes lead medieval Europeans to believe horehound provided protection from wtiches' spells. Hildegard of Bingen considered it one of the best herbs for colds. Galen was the first to recommend horehound for coughs and respiratory problems.

Hyssop : Prescribed for use in a tea for coughs, wheezing and shortness of breath and in plasters and salves for chest decongestion. Hildegard of Bingen recommended a meal of chicken stewed in hyssop and wine as a treatment for depression.

Licorice : Hippocrates recommended licorice for cough, asthma and other respiratory diseases. Hildegard of Bingen prescribed it for stomach problems.

Mint : Spearmint was the original medicinal mint and was used to aid in digestion and the treatment of gout. In Gilbertus Anglicus' Compendium of Medicine the following treatment is recommended for "stinking of the mouth":

If there be no rotten flesh, let the mouth be washed with wine that birch or mint has been soaked in. And let the gums be well rubbed with a rough linen cloth until they bleed. And let him eat marjoram, mint and parsley till they be well chewed. And let him rub well his teeth with the herbs he chewed and also his gums.

Motherwort : Used to treat heart palpitations and depression; later to stimulate contractions during childbirth.

Myrrh : Myrrh was valued primarily as an oral treatment for bleeding gums, mouth ulcers and sore throat. It was also used as an expectorant for colds and congestion.

Oregano : Used as a digestive aid, arthritis treatment, expectorant for cough, colds and chest congestion.

Parsley : Galen prescribed parsley for epilepsy and as a diuretic to treat water retention. Hildegard of Bingen recommended parsley compresses for arthritis and parsley boiled in wine for chest and heart pain.

Pennyroyal (Fleabane) : Pliny first discovered that when rubbed on the skin or strewn, penyroyal repelled fleas. He also recommended it as a cough remedy and digestive aid. Taken with honey, pennyroyal was said to cleanse the lungs and clear the chest of "all gross and thick humours".

Roses : Hippocrates recommended rose flowers mixed with oil for diseases of the uterus. Hildegard of Bingen prescribed rose hip tea as the initial treatment for many complaints including headache, dizziness and cramps. For difficulty in swallowing, Gilbertus Anglicus recommended a syrup called "honey roset" which consisted of a pound of minced roses soaked in a pound of honey over a fire.

Rosemary : In 1235 Queen Elizabeth of Hungary became paralyzed. According to legend, a hermit soaked a pound of rosemary in a gallon of wine for several days then rubbed it on her limbs, curing her. Rosemary/wine combinations became known as Queen of Hungary's Water and were used externally to treat skin problems, gout, dandruff, and prevention of baldness.

Saffron : Rare in Europe until after the Crusades, by the 14th century saffron was recommended to treat jaundice, insomnia and cancer.

Sage : Sage was considered to be something of a cure-all. Pliny prescribed it for snakebite, epilepsy, intestinal worms and chest ailments. Dioscorides recommended using sage leaves as bandages for wounds. Gilbertus Anglicus recommends the following for aching eyes:

And if it (the ache of the eyes) comes of phlegm purge it as I told in the headache and in other sicknesses of the head. And if it be of melancholy, purge it. And afterward let him be stood over a stew five or seven times, that is made of wormwood, betony, fennel, sage, flowers of thyme, chamomile flowers, melilot flowers, hock, and wild celery. All these must be soaked in wine and water together, half wine and the other half water. And let him hold his eyes and his head over the stew. And afterward take a linen cloth or cotton and wet in the water. And annoint therewith his eyes.

St. John's Wort : Early physicians prescribed St. John's Wort taken internally with wine as a cure for poisonous snakebite and externally as a treatment for burns. Christians believed that St. John's wort repelled evil spirits and burned it in bonfires on St. John's Eve to purify the air, drive away evil spirits and ensure healthy crops.

Thyme : Thyme was used as a cough remedy, digestive aid and treatment for intestinal worms. Hildegard of Bingen favoured it for skin problems. Thyme was also recommended to those who suffered from depression; they were advised to sleep on thyme-stuffed pillows. Gilbertus Anglicus recommended the following involved treatment for a soft spleen:

For softness of the spleen, if it be of cold humours give him oximel (medicine made of two parts vinegar and one of honey) to defy the humours. Afterward give him iera pigra Galieni (a decoction made of valerian, cinnamon, saffron, camel hay, hazelwort, bark of cassia and balsa, violets, wormwood, roses, gourds and alloes among other spices) to purge the humours. After the third day treat with a bath of hot herbs such as oregano, mint, horehound, thyme, rosemary and such others. And the next day let him bleed under the ankle in the underside of the left foot. And make him hot plasters of rue, celandine, and nettles to consume the humours and lay to the spleen.

Valerian : Ancient physicians recommended valerian as a diuretic, antidote to poisons, for pain relief and as a decongestant. Hildegard of Bingen prescribed it as a tranquilizer and sleeping aid.

Vervain : In the middle ages healing herbs were called simples. Vervain was prescribed so often and for so many different ailments that it became known as "simpler's joy." It was recommended for fever, tumours, blood infections, toothache, and acne among many other ailments.

Yarrow : Used to treat inflammations, to stop bleeding, as a digestive aid, for pain relief and as a mild sedative.

Medicinal Plants as a Part of Culture

It is evident that the Indian people have a tremendous passion for medicinal plants and use them for a wide range of health related applications from a common cold to memory improvement and treatment of poisonous snake bites to a cure for muscular distrophy and the enhancement of body's general immunity. In the oral traditions local communities in every ecosystem from the trans himalayas down to the coastal plains have discovered the medical uses of thousands of plants found locally in their ecosystem. India has one of the richest plant medical culture in the world. It is a culture that is of tremendous contemporary relevance because it can on one hand ensure health security to millions of people and on the other hand it can provide new and safe herbal drugs to the entire world. There are estimated to be around 25000 effective plant based formulations are available in the indigenous medical texts formulations used in folk medicine and known to rural communities all over India and around 10000 designed.

Distribution

Macro analysis of the distribution of medicinal plants show that they are distributed across diverse habitats and landscape elements. Around 70% of India's medicinal plants are found in tropical areas mostly in the various forest types spread across the Western and

Eastern ghats, the Vindhyas, Chotta Nagpur plateau, Aravalis & Himalayas. Although less than 30% of the medicinal plants are found in the temperate and alpine areas and higher altitudes they include species of high medicinal value. Macro studies show that a larger percentage of the known medicinal plant occur in the dry and moist deciduous vegetation as compared to the evergreen or temperate habitats. Analysis of habits of medicinal plants indicate that they are distributed across various habitats. One third are trees and an equal portion shrubs and the remaining one third herbs, grasses and climbers. A very small proportion of the medicinal plants are lower plants like lichens, ferns algae, etc. Majority of the medicinal plant are higher flowering plants.

Distribution of Medicinal Plants by Habits

Of the 386 families and 2200 genera in which medicinal plants are recorded, the families Asteraceae. Euphorbiacae. Laminaceae, Fabaceae, Rubiaceae., Poaceae, Acanthaceae, Rosaceae and Apiaceae share the larger proportion of medicinal plant species with the highest number of species (419) falling under Asteraceae.

About 90% of medicinal plant used by the industries are collected from the wild. While over 800 species are used in production by industry, less than 20 species of plants are under commercial cultivation. Over 70% of the plant collections involve destructive harvesting because of the use of parts like roots, bark, wood, stem and the whole plant in case of herbs. This poses a definite threat to the genetic stocks and to the diversity of medicinal plants if biodiversity is not sustainably used.

Conservation and Development of Medicinal Plants

Medicinal plants continue to be an important therapeutic aid for alleviating ailments of humankind. Search for eternal health and longevity and to seek remedy to relieve pain and discomfort prompted the early man to explore his immediate natural surrounding and tried many plants, animal products and minerals and developed a variety of therapeutic agents. Over millenia that followed the effective agents amongst them were selected by the process of trial, error, empirical reasoning and even by experimentation. These efforts have gone in history by the name discovery of `medicine'.

In many eastern cultures such as those of India, China and the Arab/Persian world this experience was systematically recorded and incorporated into regular system of medicine that refined and developed and became a part of the Materia Medica of these countries. The ancient civilization of India, China, Greece, Arab and other countries of the world developed their systems of medicine independent of each other but all of them were predominantly plant based.

But the theoretical foundation and the insights and indepth understanding on the practice of medicine that we find in Ayurveda is much superior among organized ancient systems of medicine. From history we learn that in the ancient times India was known as a place of rich natural resources, knowledge, wisdom and scholarship. People from other countries of the world as China, Cambodia, Indonesia and Baghdad used to come to the ancient universities of India like Takshila (700 BC) and Nalanda (500 BC) to learn health sciences of India, particularly `Ayurveda'. It is perhaps the oldest (6000 BC) among the organized traditional medicine. It has gone through several stages of development in its long history. It spread with Vedic, Hindu and the Buddhist cultures and reached as far as Indonesia in the east and to the west it influenced the ancient Greek who developed a similar form of medicine.

All Systems of Medicine in India functions through two social streams: Folk Stream: Comprising mostly the oral traditions practiced by the rural villages. The carriers of these traditions are millions of housewives, thousands of traditional birth attendants, bone setters, village practitioners skilled in accupressure, eye treatments, treatment of snake bites and the traditional village physicians/herbal healers, the vaidyas' or the tribal physicians.

This stream of inherited traditions are together known as Local Health Traditions (LHT). LHT represent an autonomous community supported health management system which efficiently and effectively manage the primary health care of the Indian rural mass. LHT is still alive and runs parallel to the state supported modern health care system; but its full potential is still not fully utilized and also that the great service it is rendering to the rural people go largely unnoticed because of the dominant western medicine.

Classical stream: At the second level of traditional health care system is the scientific or classical systems of medicine. This comprises of the codified and organized medicinal wisdom with sophisticated theoretical foundations and philosophical explanations expressed in classical texts like `Charka Samhita', `Sushruta samhita', 'Bhela samhita', and hundreds of other treatises including some in the regional languages covering treaties of all branches of medicine and surgery. Systems like Ayurveda, Siddha, Unani, Amchi and Tibetan, etc. are expressions of the same. Ayurveda was taught in the ancient universities in India and evolved, developed and flourished mostly among the urban centres and thus used to be a refined system of medicine.

Revival of Traditional Medicine

Today we find a renewed interest in traditional medicine. During the past decade there have been an ever increasing demand especially from developed countries for more and more drugs from plant sources. This revival of interest in plant derived drugs is mainly due to the current widespread belief that `green medicine' is safe and more dependable than the costly synthetic drug many of which have adverse side effects. This resurgence of interest in the plant based drugs have necessitated an increased demand of medicinal plants leading to over-exploitation, unsustainable harvesting and finally to the virtual decimation of several valuable plant species in the wild.

Moreover, the habitat degradation due to increased human activities (human settlements, agriculture and other developmental programmes), illegal trade in rare and endangered medicinal plants, and loss of regeneration potential of the degraded forests have further accelerated the current rate of extinction of plants particularly the medicinal plants.

Medicinal Plants Wealth of India

India is rich in medicinal plant diversity. All known types of agroclimatic, ecologic and edaphis conditions are met within India. The biogeographic position of India is so unique that all known types of ecosystems ranging from coldest place like the Nubra Valley with - 570 C, dry cold deserts of Ladakh, temperate and Alpine and subtropical regions of the North-West and trans-Himalayas, rain forests with the world's highest rainfall in Cheerapunji in Meghalaya, wet evergreen humid tropics of Western Ghats, arid and semi-arid conditions of Peninsular India, dry desert conditions of Rajasthan and Gujarat to the tidal mangroves of the Sunderban.

India is rich in all the three levels of biodiversity-such as species diversity, genetic diversity and habitat diversity. There are about 426 biomes representing different habitat diversity that gave rise to one of the richest centres in the world for plant genetic resources. The total number of flowering plant species although only 17,000, the intraspecific variability found in them make it one of the highest in the world. Out of 17,000 plants, the classic systems of medicines like Ayurveda, Siddha and Unani make use of only about 2000 plants in various formulations.

The classical traditions were prevalent in the past particularly in the urban elite society. The rural people who constitute 70 to 75% of the Indian populations live in about 5,76,000 villages located in different agroclimatic conditions. The village people have their own diverse systems of health management. While most of the common ailments were managed in the house by home remedies which included many species and condiments like pepper, ginger, turmeric, coriander, cumins, tamarind, fenagree, tulsi, etc., more complicated cases were attended by the traditional physicians who use a large number of plants from the ambient vegetations and some products of animal or mineral origin to deal with the local diseases and ailments.

These are indeed community managed systems independent of official or government system and are generally known as Local Health Tradition (LHT). The traditional village physicians of India are using about 4500 to 5000 species of plants for medicinal purpose. There is however no systematic, inventory and documentation about the folk remedies of India. There is urgent need to document this fast disappearing precious knowledge system. The oral traditions of the villagers use about 5000 plant for medicinal purposes. India is also inhabited by a large number of tribal communities who also posses a precious and unique knowledge about the use of wild plants for treating human ailments. A survey conducted by the All India Coordinated Research Project on Ethnobiology (AICRPE) during the last decade recorded over 8000 species of wild plants used by the tribals and other traditional communities in India for treating various health problems. Some interesting observations made in the study is the use of the same species found in different regions for the same ailments while some other species are used differentially.

Strategies & Priorities

The world conservation strategy (IUCN, UNEP & WWF, 1980) defines conservations as "the management of human use of the biodiversity so that it may yield the greatest sustainable benefit to present generation while maintaining its potential to meet the needs and aspirations of future generations". The above definition invokes two complementary components "conservation" and "sustainability". The primary goals of biodiversity conservation as envisaged in the World Conservation Strategy can be summarised as follows:

1. Maintenance of essential ecological processes and life support systems on which human survival and economic activities depend,

2. Preservation of species and genetic diversity and
3. Sustainable use of species and ecosystems which support millions of rural communities as well as major industries.

Medicinal plants are potential renewable natural resources. Therefore, the conservation and sustainable utilisation of medicinal plants must necessarily involve a long term, integrated, scientifically oriented action programme. This should involve the pertinent aspects of protection, preservation, maintenance, exploitation, conservation and sustainable utilization. A holistic and systematic approach envisaging interaction between social, economic and ecological systems will be a more desirable one. The most widely accepted scientific technologies of biodiversity conservation are the in-situ and ex-situ methods.

In-situ Conservation

In has been well established that the best and cost-effective way of protecting the existing biological and genetic diversity is the 'in-situ' or on the site conservation wherein a wild species or stock of a biological community is protected and preserved in its natural habitat. The prospect of such a 'ecocentric', rather than a species centred approach is that it should prevent species from becoming endangered by human activities and reduce the need for human intervention to prevent premature extinctions. Establishment of biosphere reserves, national parks, wild life sanctuaries, sacred groves and other protected areas forms examples of 'in-situ' methods of conservation.

The idea of establishing protected area network has taken a central place in all policy decision process related to biodiversity conservation at national, international and global level.

In India 4.5% of its total geographical area constitute protected area network, comprising 8 designated biospheres, 87, national parks, 447 wild life sanctuaries.

This network encompasses various biogeographic zones and biomes rich in biotic diversity, including medicinal and aromatic plants. In addition to these there area number of sacred groves in different parts of the country particularly in South, West and Eastern parts which are also active centres on in-situ conservation of medicinal plants. Such conservation area network can attribute significantly towards the conservation and sustainable management of biological resources of our country. However, experiences have amply demonstrated that in a densely populated developing country like India, where a sizeable population are living in close proximity to forests, declaring protected areas will not entirely be sufficient to ensure conservation on the fast eroding biological diversity. The success of any conservation programme vests solely on the efficient management of protected areas.

The involvement of local communities in conservation activities has now been increasingly realised. A people nature-oriented approach thus become highly imperative. This will help to generate a sense of responsibility among the local people about the values of biodiversity and the need to use it sustainably for their own prosperity and the maintenance of ecosystem resilience.

In-situ conservation of medicinal plants in India can be accomplished through the active support and participation of people who dwell in or near and around the protected forest areas. Involving the local mass in all phases of conservation programmes, such as planning, policy-decision process, implementation etc will be a significant component in achieving efficient management and utilization of medicinal plant resources. A few such in-situ conservation areas have been marked and declared as medicinal plant in-situ conservation areas on the forests of three Southern States of Kerala, Tamil Nadu and Karnataka by the joint efforts of the forest departments of these States and FRLHT, Bangalore.

Ex-situ Conservation

Conservation of medicinal plants can be accomplished by the ex-situ *i.e.* outside natural habitat by cultivating and maintaining plants in botanic gardens parks, other suitable sites, and through long term preservation of plant propagules in gene banks (seed bank, pollen bank, DNA libraries, etc.) and in plant tissue culture repositories and by cryopreservation).

Botanical gardens can play a key role in ex-situ conservation of plants, especially those facing imminent threat of extinction. Several gardens in the world are specialised in cultivation and study of medicinal plants, while some contain a special medicinal plant garden or harbour special collection of medicinal plants.

India has a network of about 140 botanical gardens which include 33 botanical gardens attached to 33 universities botany departments. But hardly 30 botanical gardens have any active programme on conservation. Tropical Botanical Gardens & Research Institute (TGBRI), located in a degraded forest region of Western Ghat mountains in Kerala has an excellent example in ex-situ conservation of plant diversity in India. The field gene bank programme launched by TBGRI from 1992-1999 is now well acclaimed as a very effective method of conservation of medicinal and aromatic plant genetic resources. This field gene bank of medicinal and aromatic plants at TBGRI, Thiruvananthapuram is essentially a blend of the ex-situ and in-situ situations.

Field Gene Bank of Medicinal Plants: The concept of establishing field gene banks of plants provide ample options for long term preservation of the genetic variability (inter-specific) of species. Field gene banks are better established in a degraded forests where efforts could be made to reforest/restock the missing species complexes, trees, shrubs, herbs, climber etc. It is indeed a recreation of a forest or rather simulation of a typical forest.

Before attempting to establish such a field gene bank it is essential to have a clear understanding of the natural ecosystem such as the spatial distribution, pattern of association *i.e.*, structure and functional dynamics of the species in question. After undertaking a in depth study on the natural distribution pattern of the medicinal plants and the associated floristic elements - including their micro-ecological niche, a well planned action programme of recreating the same in a degraded forest area or place close to the species found in nature can be attempted.

TBGRI has accomplished this task of simulating the nature while establishing the field gene bank of medicinal and aromatic plants under the G-15-GBMAP sponsored by DBT, Government of India. TBGRI experience now provide ample opportunity to repeat the same elsewhere in the country.

Identification of the keystone species and umbrella species are very important in this methods. After planting the keystone and umbrella species, other species complex which include the medicinal aromatic plants in question have to be introduced. The sampling and selection of samples for introduction have to be highly knowledge and science intensive. To capture the maximum possible genetic diversity of the target species it is extremely important to collect all valuable information such as morphological variants, chemical variants or genetic variants or chemical screening of the population of the targeted species by using the latest methods and tools.

The field gene bank of TBGRI has covered 30,000 accessions of 250 medicinal and aromatic plant species which include 100 endemic, rare and endangered medicinal and aromatic plants of the tropical region of India. A broad spectrum of the genetic diversity of these species were captured and introduced in this gene bank which covered morphotypes, cytotypes and chemotype and the number of samples from each species varied from 50-1000 plants.

Some Economically Important Aromatic Plants

Tongkat Ali (Eurycoma longifolia, Simaroubaceae)

Tongkat Ali is also known as tunjang bumi, pasak bumi and penawar pahit. It has long been used in traditional medicine, especially by the Malays and Orang Asli. The roots are boiled and used as an aphrodisiac and tonic for men; it is also used for treating malaria and fever. This crop grows well in deep sandy loams mixed with plenty of organic matter. The roots can be harvested at the age of at least five years.

Kacip Fatimah (Labisia Pumila, Myrsinaceae)

Kacip Fatimah is a small, slighty woody herbaceous plant, which can be found in forest throughout Malaysia from the sea level up to about 150m altitude. A decoction of the roots is used for post-partum treatment, gonorrhoea and rheumatism. This plant can be propagated through seed and vegetative cuttings and can be harvested at the age of 7-8 months.

Mengkudu (Morinda Citrifolia, Rubicieae)

Mengkudu is a small tree which can grow up to 6m tall with few spreading branches. It grows well on clay loam in full sunlight or in some shade. It flowers and fruits all year round and can be propagated by stem cutting or seeds.

The roots are boiled and the resultant decoction drunk to encourage the onset of menstruation. The leaves are heated over a small fire and applied to the chest to relieve cough or to be abdomen for mothers after childbirth.

Kadok (Piper Sarmentosum, Piperaceae)

Piper sarmentosum, known as daun kaduk is very popular in Malaysia and is often mistaken for its cousin Piper betel leaf plant. A decoction of the boiled leaves has been known to be effective in treating coughs, flu, rheumatism, pleurisy and lumbago. The root is a remedy for toothache and may be made into a wash for fungoid dermatitis on the feet. The leaves are also reported to contain antioxidant property.

Serai Wangi (Cymbopogon Nardus, Graminiae)

Serai Wangi is the perennial crop which grows well in sandy loam in full sun and establishes itself quickly into a bush. Infusion of the leaves is sometimes used in herbal bath for mothers to regain health after childbirth. The essential oil obtained from this herb is also used extensively in cosmetics.

Daun Kesom (Polygonum Minus, Polygonaceae)

Daun Kesom is a bushy herb to 50cm tall which can be found growing in shallow ditches and wet places. It thrives on sandy loam in full sun and flowers upon maturity. It can be propagated from stem-cutting or seeds and is often planted for flavouring in cooking due to its strong aroma. Atrong docoction of the fresh herb is taken foe indigestion and as a remedy for stomach pains.

Akar Cerita (Andrographis Paniculata, Acanthacea)

Akar Cerita is known for its incredibly bitter taste so much so it is also called 'hempedu bumi' or 'bile of the earth'. It prefers rich loamy soils with some shade. It flowers frequently and can be propagated from stem-cuttings or seeds. A decoction of the leaves is often taken orally to cure diabetes and to reduce high blood pressure. A leaf poultice is applied topically to relieve itchy skin and insect bites. This herb is also said to be useful as a liver tonic to help detoxify toxins in our body.

Pegaga (Centella Asiatica, Apiaceae)

Pegaga, a popular 'ulam' (salad) is a creeping herb with a long stolen. It grows well in the open partially shaded habitats and is often planted in small scale for its medicinal properties. A decoction of pegaga have been used to treat skin diseases, hypertension and to improve blood circulation. The stolon with stems and roots may be used for cultivation. The plant can be propagated vegetatively using the stolon. It is ready for harvesting six months after planting whereby clumps of plants can be easily dug up using small spade. Pegaga is also a popular ingredient in cosmetic products.

Medicinal and Aromatic Plants

Traditionally, raw materials of most medicinal and aromatic plants have been sourced from natural forests. Continuous extraction from this source without concerted efforts on replacement through replanting has inevitably led to the depletion of these important raw materials. One of the key determinant of the future development of the medicinal and aromatic plants industries in this country is the sustainable supply of the raw materials. For continuous and sustainable supply of the raw materials, some forms of planting are deemed necessary.

A major limitation in the widescale planting of potentially high economic value crops such as medicinal and aromatic plants is the issue of the land availability. Land for the planting of any crop, used to be available in abundance but is currently getting scarce. Futher more, because environmental consideration, clearing of forests for the purpose of crop cultivation is currently not encouraged. There is a tremendous pressure to conserve our forests. In addition, a lot of our natural forests have been gazetted as permanent forests.

Even in new plantings, with the good price of palm oil, the focus is more on the planting of this crop. In addition, monoculture planting of medicinal and aromatic plants, although of potentially high value, may involve some form of biological and economic risks. With this scenario, it is therefore imperative that alternative forms of planting of medicinal and aromatic plants, are sought and considered.

One option available is on agroforestry (Mahmud, 1997a), a combining agricultural crops such as oil palm with medicinal and aromatic plants (Mahmud, 1997b, c) on the same piece of land. Under the concept of maximum land utilization and the need for diversification and alleviating potential risks in planting, medicinal and aromatic plants providing added value to the land. In any agroforestry planting system to be adopted, due consideration should be given to minimizing possible competitive effects between the various component species and the provision of conducive environments for the proper establishment and growth of all component species.

In view of this, various factors (Mahmud, 1997a) have to be taken into consideration in considering the integration of medicinal and aromatic plants with oil palm and these include:

i. Growth habits of forest species in terms of growth rates, crown shape and size etc.
ii. Growth requirements for light, moisture, nutrients, space etc., of all component species.
iii. Duration of growth
iv. Topography, either flat, undulating or hilly terrain.
v. Planting direction in terms of maximizing light transmission and capture.

Taking the above factors into consideration, the following are some illustrations and interim proposals on the integration of selected medicinal and aromatic plants with oil palm in an agroforestry system of planting.

Integration of Tongkat Ali with Oil Palm

Tongkat Ali which is targeted for harvesting at 4 to 6 years after planting, has a monopodial growth habit with very limited branching. It can be possibly be integrated with oil palm without having to modify the existing planting system of oil palm. In view of its simple growth habit, the Tongkat Ali can be planted in a single row at a distance of three metres apart in the interrow spaces.

The planting can be done either in every palm rows or in alternate palms rows with the density of the Tongkat Ali in varying with the planting system. Since the Tongkat Ali may require initial shade in its early stages of establishment, the Tongkat Ali can be introduced at one, two or three years after the planting of oil palm with the expanding oil palm fronds providing the initial shade. The price of dried root of Tongkat Ali is between RM 30.00 - RM 40.00 per kg.

Integration of Kacip Fatimah or Serai Wangi with Oil Palm

Kacip Fatimah is a small slightly woody herbaceous plant which is harvested for its roots at he age of 7 - 8 months after germination. Serai Wangi belongs to the grass family which survives as clumps with the stems harvested at eight months after planting. Both these plants can be integrated with oil palm, utilising the abundant space in between the palms.

The Serai Wangi can be introduced at the same time as planting of the oil palm and the Kacip Fatimah, because of the need for initial shade, at 0.5, 1 or 2 years after oil palm planting. Because of the shorter growth duration, high density planting of Kacip Fatimah and Serai Wangi can be practised when these plants are interplanted with oil palm.

To avoid possible competition, the planting should be confined in the interrow spaces at spesific distances from the palm rows. Depending on when these crops are introduced, as much as 2 to 3 harvests of the Kacip Fatimah and Serai Wangi can be achieved with this form of planting. The price of the dried root of Kacip Fatimah and fresh leaves of Serai Wangi is RM 30.00- RM 50.00 per kg and RM 1.20, respectively.

4

The Organic Production of Essential Oils

Organic essential oils are derived from plants that have been grown without the use of pesticides, on land that has been certified by an authorised regulatory agent such as ECOCERT or the Soil Association. Today, many aromatherapists and nurses prefer to use organic essential oils in their clinical treatments because they believe they have more healing power and vitality than conventional essential oils. There is also the issue of pesticide residues to be considered too, since they have far-reaching effects for both the environment, and our bodies. Every one of our Organic Essential Oils has been analytically tested for purity and certified under one of the following official regulatory agencies:

- The Soil Association
- Nature et Progres
- ECOCERT
- Qualite-France SA
- Agrobio.

Quinessence organic essential oils bring you unrivalled value for money because in most instances we have purchase them directly from the farms where they are produced, thereby cutting out the middle-man. We then pass these savings along to you, and in many cases our organic essential oils are not a great deal more expensive than many of our conventionally produced essential oils.

Further to the organic essential oils we purchase from around the world, an increasing selection of our Certified Organic essential oils are now distilled on-site from medicinal herbs and plants grown on a 500 acre farm in the United Kingdom. Whilst our preference is for organically produced essential oils, we do accept that it has yet to be proven scientifically they are any more effective than their conventionally produced counterparts. But even if it was proven there is no difference between them, we would still not change our view on this matter. There is much more to this subject than just efficacy.

By purchasing organic essential oils from growers who use traditional farming practices, we are all contributing towards a more sustainable ecological environment for the future. Surely this is a sensible and worthwhile investment for our forthcoming generations? A large amount of agricultural land has already been lost due to soil erosion, and in many places the overuse of aggressive agrochemicals has destroyed the delicate balance between wildlife and its natural habitat.

Discover more about the sources of Quinessence organic pure essential oils and the importance of buying organic by browsing other pages under this category. Organic essential oils are a speciality here at Eden Botanicals. We continue to stock more and more organic aromatherapy essential oils – approximately 30 currently – and offer them at wholesale pricing. And as the demand for high quality, certified organic oils increases, we will keep adding more organic oils to our stocks. We are finding that the artesian distillers, as well as larger distilleries of organic essential oils for the aromatherapy market continue to increase the types of organic oils being produced. With the increase in production, we have also noticed that the quality of the organically grown essential oils is on the rise as the new distilleries become more experience in the art and science of distilling essential oils.

And, we also might as well mention that we also carry organic carrier oils such as Jojoba oil and high-oleic acid Sunflower oil – perfect for blending organic essential oil.

So, what is an organic essential oil? Just like food – essential oils are agricultural products and the rules as to what constitutes an organic agricultural product are made and regulated by governmental agencies such as the Department of Agriculture (USDA) in the USA. This program is called the national organic program (NOP). In Europe, Japan, Australia and elsewhere the programs are similar in that third party certifying agencies make sure that each organic farm (and processing or handling facility) meets the government regulation. The actual farm and facilities are certified (rather than the product itself), but the products that are grown and processed at certified facilities are what we know as certified organic products (in this case essential oils). We would also like to tell you about one of the myths surrounding "organic" essential oils found at numerous essential oil web sites. Just like food products, an essential oil is either certified organic or it is not (although the national organic program allows for either 100% organic or 95% or more organic). If the product is not certified - it is not organic (according to the government). And what this really means is that you cannot sell products as organic unless they are certified organic. This does not mean that you cannot grow crops organically unless they are certified. It simply means for it to be truly organic and sold as such it has to be certified according to the various rules in each country.

Another myth is that an organic essential oil is better than a non-organic essential oil of the same species. What makes and essential oil better (in aroma or for aromatherapy use) than another is not how the plant material was grown (although this certainly plays a big part). What makes a great essential oil is the sum of many factors such as the seeds or cuttings used to start the plants, the growing methods, harvesting methods, distillation process, and after care (post-distillation). So, while an essential oil may be certified organically grown, it may have been distilled poorly, or not cared for prior to distilling or after distilling. In fact, we have sampled many certified organic essential oils that were of a lower quality that corresponding oils that were grown conventionally. Fortunately we are experiencing fewer organic oils of low quality and far more superior quality organic essential oils. We believe this to be due to new organic farmers gaining experience and new distillers who start out distilling organic crops gain experience.

There are a couple of reasons one my wish to purchase organic essential oils. The first is that you may want to support organic agriculture, organic farmers and sustainable practices. This is a main reason that we carry and use organic essential oils. The less pesticides, herbicides, chemical fertilizers used on our planet the better life will be for humanity and the natural environment. Another reason to purchase and use organic essential oils is because you are making aromatherapy, body care or perfumery products and would like to either say your product is organic or you would like to label it and have it certified as organic.

Whatever your reason is for using organic essential oils we hope that you will continue to learn about essential oils and use them in safe and responsible way. We also hope that you enjoy browsing our web site and learning more about aromatherapy, natural perfumery and organic essential oils. Agriculture, our most important global industry, provides the world's population (currently three billion) with food from plants and animals. Farmers grow plants and animals to harvest as foodstuffs and are limited, mostly, by the local weather, pests (eg, rabbits, pigeons, rats, slugs, snails, animal parasites, soil nematodes, insects, bacteria, fungi and viruses), weeds, plant and animal diseases. This means that in the UK, in our cultivated fields, pastures and orchards (assuming that we can control pests and diseases), we can only grow plants (as crops) which can withstand low temperatures and frost.

However, despite the limitations of the climate, UK farmers have achieved outstanding success in providing us with healthy, nutritious, available and cheap food (meat, dairy products, eggs, cereals, oil-seeds, potatoes, vegetables, fruit). In fact, in many ways, farmers are victims of their own success and have produced too much too quickly for a consumer turning increasingly towards different food (curry, pasta, rice) and away from animal products. On top of this a very sinister disease emerged in our national dairy herds, appropriately called "mad-cow", which has terrified the European community.

Figure : *Electron micrograph of the peppermint oil gland on the leaf. The arrows indicate that the gland is filling with oil.*

Other glands are visible which do not contain an oil.

Figure : *A field of lavender cultivated on Merseyside after two years.*

The problems of over-production, food mountains, infected livestock, anxious consumers and bankrupt farmers have led to a real national disaster which has to be resolved because in forty years time, the world population will have tripled to over nine billion people, all needing food, energy and materials.

It must also concern us all that our very valuable agricultural land is disappearing, silently and secretly. This land is lost forever because the top soil disappears. This layer has taken hundreds of years to evolve into a mixture of sand, silt, clay, organic matter, animals and microbes which provides the essential and critical fertile environment for the plants to grow. On Merseyside, over a ten year period, two hundred hectares of very productive and naturally fertile arable land has disappeared each year.

Farmers need alternative, non-food, sustainable crops for their land and many industries (pharmaceutical, complementary health, flavour and fragrance) increasingly need to source high-quality plants and plant products. Coupled with this, the consumer (aromatherapist, medical herbalists, pharmacist) is aware that good quality, botanically authenticated, organically grown plants offer a superior product.

Research into temperate-climate essential oil plants (peppermint, lavender, Roman chamomile, German chamomile, sage, clary sage, angelica, helychrysum, dill, etc.) as crops began a number of years ago at Liverpool's John Moores University. The results indicated the potential of these plants as alternative crops for the farmers, as the plants produced essential oils of excellent quality to supply the increasingly demanding market place. As a result, a unique partnership between academics, researchers, farmers and end-users (aromatherapists, medical herbalists) was initiated in 1995 to develop new commercial crops for the UK and to grow these crops using organic systems of husbandry, as 1 well as traditional methods.

Strictly speaking, the term "organic" relates to the unique characteristics of plants and animals, and their products, because of their carbon basis. The term also refers to a class of chemical compounds that are formed from carbon. Carbon, in the form of carbon dioxide, enters the plant via tiny microscopic holes, most often found in the leaf, the stomata. From there it enters the chloroplast and, when the energy of the sun contacts the chlorophyll, a tiny electrical current is initiated. Water and carbon dioxide molecules are broken down and rearranged to produce sugar (carbohydrate) and oxygen via a very complex process called photosynthesis. Many tonnes of carbon are taken from the atmosphere and "fixed" by plants in this way, and animals acquire the carbon by eating the plants. It is a very intricate process and vital to life. The term "organic" is also used to describe a particular method of growing plants (and animals). Although the term has become familiar, and certainly very popular, most people only have a vague idea of its meaning and range. We do need to be precise about these terms though, and clear about our definition of desired product quality.

Organic farming is the outcome of thinking and practice since the early twentieth century and involves a variety of alternative methods of agricultural production. Although a full description of the history and principles of organic production, the interested reader is directed to a comprehensive publication on the principles of organic farming and towards a book which is compulsory reading for all those interested in the environment, and the effects of the indiscriminate use of chemicals by various industries.

Organic farming can briefly be described as using methods of crop (and animal) husbandry which coexist with, rather than dominate, natural systems, sustain and build soil fertility, minimise pollution and damage to the environment and ensure the ethical treatment of animals. This serves to protect and enhance the farm environment with particular regard to conservation and wildlife by using sustainable crop rotations which avoid the use of fertilisers in the form of soluble mineral salts. The use of agro-chemical pesticides is prohibited. Animal husbandry techniques are used which meet the animal's physiological, behavioural and health needs. Organic farming pursues a number of aims such as the production of quality agricultural products which contain no chemical residues, do not pollute the soil and groundwater with pesticides and make full use of natural, local and renewable resources.

The consumer needs to be aware of a number of issues regarding organic, and so-called organic production. First of all, many farmers already follow the above criteria when growing their plants and animals. Failure to follow the above procedures does not mean that farmers are deliberately poisoning the consumer, the land and their animals. Many herbal products and essential oils have appeared in the marketplace labelled "organic" without detailing methods of production and can be misleading for the consumer. We need to be aware that legislative criteria exists in the UK for producing plants and their products organically and to be clear about what this means for farmers and growers.

Medicinal and aromatic plants (many called herbs) synthesise and accumulate oils in discrete structures (known as glands, or trichomes) often distributed on the leaves and flowers. Although we don't know why a plant should make these structures and fill them with oil, it seems obvious that, during evolution (bear in mind that plants were on the globe before animals), plants had to defend themselves from parasites and predators such as fungi and insects, and subsequently, grazing animals. Both glands look like weapons. The plant needs weapons as a defensive strategy as they are lodged in the soil by their roots and can't escape from danger.

We have examined the oils, identified their components and tested the oils against bacteria and fungi. They are very biologically active and can, in fact, act as pesticides – bactericides, fungicides, insecticides and, in some cases, herbicides. But, is this surprising? Why should a plant invest all that genetic material (DNA, enzymes, photosynthate) into producing something which is useless? Plants need to survive and therefore need to be competitive. Studies on many plants reveal them to be ruthless in their determination to survive and leave their own offspring. Plants have more genetic material than most animals and can count time very efficiently as well as measure and react to, minute changes in day-length and temperature.

Many cultivated agricultural crops have had their defensive strategies bred out of them in favour of more valuable characteristics such as yield. Medicinal and aromatic plants, as crops, are really wild plants as they haven't been selected and bred for optimum cultivation in a particular environment. In practical terms, this means that they don't necessarily respond to fertilisers. In reality, many produce higher yields of oil due to environmental stresses. Since they have not been cultivated on a large scale in the UK, they have not prompted the upsurge of crop pests and diseases and this also may be due to the properties of the oil. In 1996, major UK arable farmers began the cultivation of medicinal and aromatic plants on their farms on Merseyside. They undertook to cultivate unknown perennial plants (peppermint, Roman chamomile, lavender, sage) to produce an essential oil without knowing how to get the oil out of the plant. Added to this they agreed to use organic systems of husbandry. In the UK, in order to comply with legislative criteria regarding organic cultivation, farmers must comply with European Council Regulation (EEC) No 2092/91, which came into force on July 22nd 1991, and which lays down the main principles for organic production and the rules that must be followed for the processing and sale of organic products. The advantage of these legal instruments is to formalise recognition of the organic sector, to lay down common rules for operators who regard themselves as belonging to it, and to guarantee consumers with a means of unmistakably identifying genuine organic produce, thus effectively eliminating the abuses previously frequent in the sector.

Figure : *Close-up of German chamomile at harvest. The organically grown plants are full of oil in the flower head.*

In reality, this has taken the farmers three years. During this time they have taken major areas of their land out of conventional husbandry, they have put this land into conversion prior to full-scale organic production and they have applied for status with the Soil Association. They are cultivating medicinal and aromatic plants in a completely chemical-free environment. Of course, this means that they are subject to inspection by the Soil Association and must sign a contract with them agreeing to carry out operations in accordance with their defined standards.

What has been achieved for the farmers and for the consumer and what does the future hold for the industries involved? During the last few years a dedicated team has been created to produce novel, high-quality, organically grown medicinal and aromatic plants for the production of essential oils in the UK. Fields of lavender, peppermint, Roman chamomile and German chamomile have been sown and planted. As the crops have been growing in the Lancashire fields, designers and engineers have constructed on-farm hydro-distillation apparatus for the commercial extraction of the oils from the plants. Scientists have left their traditional university laboratories and relocated to the farms where the farmers have provided space in redundant buildings for elaborate equipment to measure and monitor their crops and the essential oils they produce (via Gas Chromatography).

The farmers have produced some of the finest oils the market place has ever seen in a remarkably short space of time, using published systems of organic husbandry, and have therefore produced a unique UK product. Alongside the farmers, major end-users (including aromatherapists) have been appraising the oils including the use of organoleptic techniques (taste and smell), scrutinising the Gas chromatography / Mass spectroscopy data (provided by the on-farm laboratory) which analyses the component identification data, and visiting the fields of plants. The farmers are committed to growing these new, high-value crops and achieving maximum quality for the consumer, as well as providing an excellent value-for-money product. A whole new UK industry is being created and will bring with it numerous benefits for all those concerned.

There are a vast number of essential oil products on the market today. Additionally, there are also large discrepancies in the prices of these essential oils. Producing Certified PBTG essential oils is very costly! It requires thousands of pounds of raw material to produce small amounts of quality oil. For example: it takes approximately 5,000 pounds of rose petals to produce 5ML of rose essential oil. If you purchase inexpensive rose oil, rest assured you are not purchasing a therapeutic grade Category A essential oil. Apply that same rule of thumb to all therapeutic grade essential oils. There are several companies that buy perfume or food grade products, advertise them as therapeutic grade and charge a premium.

Anyone that has discovered the world or aromatherapy and the use of essential oil understands that acquiring the right oil is vital. Inferior quality, improperly produced or modified (synthetic) oils, will not produce therapeutic results and could possibly be harmful. The key to producing a therapeutic grade essential oil is to preserve as many of the delicate aromatic compounds during production as possible. As you begin to understand the benefits of essential oils in the area of personal, holistic health and therapy, you will begin to appreciate the necessity and cost benefit of obtaining Certified PBTG essential oils.

The following physical tests and harvest considerations are conducted on all Certified PBTG essential oil products.

- GC Results
- Season and Year of Distillation
- Odor
- Colour
- Refractive Index
- Specific Gravity

- Optical rotation
- Closed-cup flashpoint.

Grading Categories for Essential Oils

Category A+: Certified NOP Organic Essential Oil

Category A+ Certified NOP organic essential oils are always made from costly, organically grown plants and use extravagant distillation processes during extraction to meet NOP organic grade standards. These oils are generally not produced due to the extreme expense of the planting, cultivation, harvesting, and extraction. If you were to locate true NOP organic essential oils, the cost to the average consumer would be out of reach.

For example, some brands of Sacred Frankincense sell for $10,000 for 5ML. Unaffordable? Absolutely, but these are true NOP Organic. True NOP Organic essential oils fall into category A+ 10 range and are very rare.

Category A: Therapeutic Grade Essential Oil

Category A essential oils are pure therapeutic grade quality and are generally, upon availability, made from organically grown plants and are extracted at temperature through steam distillation. These essential oils are costly because the ingredients that combine to create the essential oil are of the purist grade, creating a highly therapeutic benefit. All Certified PBTG essential oils fall into Category A ranging from 9.0 to 9.9.

Category B: Food Grade Essential Oils

Category B essential oils often contain pesticides, fertilizers, and synthetic extenders called carrier oils. These additives are used to bring down the cost of production and dilute the potency of the essential oil. Essential oils are considered food grade due to the adulteration they have with synthetic extenders (carrier oils). Unfortunately, Category B oils are often marketed as therapeutic grade essential oils, promising results that they are unable to produce. All Category B essential oils fall into a range of 5 to 8.9

Category C: Perfume Grade Essential Oils

Category C perfume grade essential oils may contain some similar ingredients to those of Category B food grade essential oils. Category C oils often contain solvents that are used to create higher yield per harvest. This maximum yield per harvest degrades the effectiveness of the essential oil and no therapeutic benefits are found. These similar chemicals are often found in perfumes and some store bought fragrances. Unfortunately, there are companies advertising perfume grade essential oils as though they are of Category A value. All Category C perfume grade essential oils fall into a range of 1 to 4.9.

The Cleopatra of Essential Oils

Rose oil is an essential oil found in many quality Skincare products and is used in the Natural therapies industries. It is expensive oil made from Rosa damascena commonly referred to as the Damask Rose or Moroccan Rose. Rosa centifolia is also used in some countries to produce rose oil. Rose hip seed oil is made from the Rosa rubignosa. Rosa rubignosa grows wild in the Southern Andes.

The most prized Rose Oil is produced from the petals. Rosehip oil is made from the seed capsules. Different extraction methods are required for each oil type. The manufacture of Rose oil is a very labour intensive task. The petals are collected early in the morning of the same day it is processed. Steam distillation is the most common form of extraction of the oil of rose petals. Rose hip oil is a pressed oil made by pressing the seeds to extract the oil. Rose hip oil is high in Vitamin A (retinol) and Vitamin C. Rose hip oil is also high in the Fatty acids Omega-6 and Omega-3. Rose hip oil is commonly found in many skin care products.

Rosehip Seed Oil is common in skin care products because of its Vitamin A content. Vitamin A helps to delay the effects of skin aging and assists with cell regeneration. It also promotes the increase of collagen and elastin levels.

Rosehip oil also contains Vitamin E. Rose hip oil is safe to use by anyone of any age. All types of skin can benefit from Rosehip oil.

When used as a preventative measure Rose hip oil can reduce premature aging and when applied daily may reverse the damage caused by wrinkling. When applied regularly, Rosehip oil will make the skin appear healthier.

Research has shown Rose hip oil to be effective with:

- Scarring (from wounds, Chickenpox, eczema, acne and psoriasis), Burns, Dermatitis and Skin damage caused by Sun Exposure
- Premature aging, Fine lines and Wrinkles and Stretch marks
- Hyper-pigmentation, Skin elasticity and resilience
- Dry Skin and Skin tone.

The application of Rosehip oil is by gentle massage into the affected area of skin. Depending on the size of the area being treated it can be applied by the palm of the hand but is more generally applied with the tips of the fingers to avoid wastage. Estimates show, over one million blooms are used to make one kilo of the highly prized Rose oil. No wonder it is referred to as the Cleopatra of rose oils.

Aromatic plants have been used medicinally and therapeutically throughout history. Many common plants have medicinal properties that have been applied in folk medicine since ancient times and are still widely used today. We are now just beginning to research and document the properties of the substances contained within aromatic plants, as well as recognising their abilities to aid in the healing process. But unlike herbalism, aromatherapy draws on the healing powers of plants that are found only in their essential oils. Depending on the plant involved, these aromatic, dynamic, healing essential oils may be found inside the roots, wood, leaves, flowers or fruit.

Essential oils are extracted from plants chiefly through steam distillation (roots, wood, leaves and flowers) or cold-pressing/expression (citrus oils from peel). You can make your own orange essential oil by squeezing the peel. Try it! It smells fresh and delicious. Orange peel contains large quantities of oil, and is easy to extract, making it one the cheapest oils to buy. In contrast to this is the rose, which has very little oil in the flowers, making it the most expensive oil. It takes between 3,000 and 5,000 kg of flowers (more than one million flowers!) to produce a single kilogram of rose oil. A collector usually gathers 25 kg of blossoms a day. Lavender flowers have a more accessible oil; 3 kilograms of essential oil can be harvested from 100 kilograms of flowers.

Essential oils have been described as the blood of plants, or the vital energy. They are very complex compounds that may contain several hundred different natural chemicals. These are very powerful and concentrated and need only be used sparingly and in small quantities. In many countries, essential oils are included in the national pharmacopoeia. In France aromatherapy is incorporated into mainstream medicine, and some essential oils are regulated as prescription drugs, and thus administered by a physician. There, the use of the antiseptic, antiviral, antifungal, and antibacterial properties of oils in the control of infections is emphasised.

Aromatherapy in History

We know that the Ancient Greeks, Egyptians, Chinese, Druids, Celts and many other tribes in Africa, America and Australia have made use of various aromatic plants in the ritual, medicinal, scientific and personal aspects of their lives over a long period, dating back to beyond 2000BC. Yet it wasn't until the dawn of the twentieth century when the French perfumer and chemist Gattefosse published results of his experiments with essential oils and coined the term 'aromatherapy' that the movement as we know it today, really began.

It was taken up by the physician Jean Valnet MD and biochemist Marguerite Maury and gained momentum during World War 1 when essential oils provided ready solutions for healing burned and wounded soldiers. Almost a century on, as the world wide web speeds information around the globe and science puts proof to old beliefs, aromatherapy as a practice is growing rapidly in reality and stature.

'Essential oils are one of the great untapped resources of the world. Here we have a system of natural help that is far more than a system of medicine, that can prevent illness and alleviate symptoms.' 'These extremely complex precious liquids are extracted from very specific species of plant life and are in harmony with people and planet alike'. 'By taking essential oils into our lives, we find a way to provide family and home with ? protection and pleasure... without polluting ourselves or our environment with chemicals.'

Why certified organic essential oils? Organic systems work in harmony with nature, keeping harmful chemicals out of our land, water and air, creating a healthy environment rich in wildlife, woodlands and nutrients. Organic standards place great emphasis on building and maintaining healthy soil and high vitality crops.

Simply stated, organic products are grown without the use of pesticides, synthetic fertilizers, sewage sludge, genetically modified organisms, or ionising radiation. The USDA National Organic Program (NOP) defines organic as follows: Organic food is produced by farmers who emphasize the use of renewable resources and the conservation of soil and water to enhance environmental quality for future generations. Organic plants are produced without using most conventional pesticides; fertilizers made with synthetic ingredients or sewage sludge; bioengineering; or ionising radiation.

Before a product can be labelled "organic," a Government-approved certifier inspects the farm where the food is grown to make sure the farmer is following all the rules necessary to meet USDA organic standards. Only certified organic products can bear the seal of approval. The organic seal assures consumers of the quality and integrity of organic products. Organic-certified operations must have an organic system plan and records that verify compliance with that plan. Operators are inspected annually in addition there are random checks to assure standards are being met.

Why does organic cost more? Consider these facts:

1. Organic farmers don?t receive federal subsidies like conventional farmers do. Therefore, the price of organic food reflects the true cost of growing.
2. The price of conventional food does not reflect the cost of environmental cleanups that we pay for through our tax dollars.
3. Organic farming is more labour and management intensive.
4. Organic farms are usually smaller than conventional farms and so do not benefit from the economies of scale that larger growers get.

How do oils work in the body? Tiny molecules of essential oils are taken into the body in two ways by osmosis (through the skin) and by olfaction (breathing them in).

Essential Oils and Osmosis (External Application)

When essential oils (dissolved in a carrier oil) are applied externally via massage, the tiny molecules are absorbed through the skin and reach small blood vessels. They are then carried to muscle tissue and joints via the blood stream reaching all the tissues and organs. The oils are then excreted through the kidneys and bladder, skin, and exhaled through the lungs.

Essential Oils and Olfaction (Inhalation)

When essential oils are inhaled, the molecules are absorbed directly into the bloodstream via the lungs, which affects the entire respiratory system, and are absorbed by the olfactory nerves through the nose, where they travel directly to the limbic system that deals with integration and expression of feelings, learning, memory, emotions and physical drives. Once they reach the limbic system, they trigger the release of neuro-chemicals which may be sedative, relaxing, stimulating or euphoric in effect.

Today, there are about 300 oils in professional use around the world. Increasingly, commercial enterprises such as hospitals and medical centres, departments and retail stores, aged care facilities, airlines and other organisations are realising the many of benefits to be gained through judiciously dispersed essential oil blends like lemon (proven to reduce clerical errors) and vanilla (proven to make shoppers linger longer).

For the 'ordinary' user, essential oils open up a plethora of safer, 'green' choices in a world that's daily becoming more chemically threatening. Most experts recommend that you start with three or four common oils, learning all you can about them, using them and noting their effects until you become completely familiar with what they can do for you. And 'for you' is an important point since the effects of essential oils must always be intensely personal and subjective. Then, move forward in increments of say - two new essential oils until you have command and a thorough knowledge of all the oils that you need to make your family happy and healthy.

How to use Essential Oils

Essential oils are used in a number of ways that have developed over time in different areas around the world. Some oils have a recorded use of more than 4,000 years. The modes of application of aromatherapy include:

- Aerial diffusion: for environmental fragrance or disinfection.
- Direct inhalation: for respiratory disinfection, decongestion, expectoration as well as psychological effect.
- Topical applications: for general massage, baths, compresses, and therapeutic skin care.

Vapouriser/Room Burners/Electric Diffusers

The most common way to create a beautiful atmosphere or disinfect a room is to add essential oils to a vapouriser which can be made from different materials - ceramic, terra-cotta, metal or glass - with two separate parts - the top one for water and essential oils and the bottom part for housing a tea-light candle to provide the gentle heat.

When purchasing a vapouriser opt for one with a large top reservoir so you don't need to constantly top it up. The general rule is about 8-15 drops of essential oil in the water in the top reservoir, depending on the size of reservoir and the size of the room. Ensure your vapouriser is placed on a heatproof stand away from draughts. As the water heats, the essential oils will be diffused. Keep an eye on the bowl to ensure the water doesn't totally evapourate. If it evapourates before the essential oil has been vapourised, you could end up with a spatter. And never leave the house with a candle burning! If the idea of burning candles is a worry to you, you can use electrically powered aromatherapy diffuser.

Steam Inhalation

To inhale steam directly, you need a large heatproof bowl and a thick bath towel. Place your bowl onto a solid surface such as a table and fill it with boiling or almost boiling water. Begin by adding 3-6 drops of your chosen essential oil. As that dissolves, add two more drops and then again to a total of 6 drops. Seat yourself safely and comfortably; lean over the bowl and use the towel to seal off the vapours. Breathe in slowly and quietly for between 1-5 minutes. You may want to shut your eyes. This process of taking the essential oil directly into your nose, throat and chest has strong antibacterial, antiviral and soothing effects. If you'd like to gain an extra benefit from this procedure, splash your face with cold water afterwards for a skin tingling mini-sauna.

Bath

Baths are one of the easiest and most pleasurable ways of using essential oils. An aromatic bath can refresh you body, mind and spirit. Add 10 drops in a tablespoon of carrier oil into your warm to very warm bath, swish the water around to mix the oils in. For maximum benefit soak for at least 10 minutes. After your bath make up a lovely massage blend and massage all over your body to prolong the benefits of the bath and nourish and moisturise your skin.

Room Freshener

Essential oils are an ideal way to disinfect or fragrance any room. Your whole house doesn't have to smell like bathroom spray. Not only will your room be fragranced beautifully but you will also benefit from the therapeutic properties of the oils. Simply add 20 drops of your favourite essential oils into a MiEnviron spray bottle half-filled with water, shake vigorously and you're away! Can be used to spray on bedsheets and pillows but make sure not to spray on polished furniture, as some essential oils can damage the polished surface. Store the bottle in the fridge.

Household

There are lots of ways to germ proof and fragrance your home with essential oils. A few drops of something you love in the bag (or on the filter) of your vacuum cleaner will diffuse fragrance as you clean. Similarly, you can add a trace of scent to dusting cloths, clothes washers, floor cleaners and polishes. Furniture polish is not the same without lavender. And the list goes on.

Compress

Fill a bowl with warm water and add 2-6 drops of the required essential oil. Stir well. Then take a soft, clean cloth, soak it throughly, wring and place gently but firmly on the affected body part. Repeat the procedure until the discomfort is relieved. Be very careful not to allow the fluid into the eyes, nose or mouth.

Skin Care

Many essential oils have lots of uses when applied directly to the body. But oils are very highly concentrated, so it's never wise to put them on the skin without first mixing into a moisturiser or oil. This is one case where a heavier oil will come into its own. If you're caught short, explore your pantry for olive, grapeseed or safflower oils. Use 6-8 drops of your selected essential oil into two teaspoons of your carrier oil. There are two exceptions to the rule of 'don't apply direct.' These are lavender which can be used for cuts and minor burns and tea tree oil which can be applied topically to insect bites, cuts and scrapes.

Bath Salts

Add 15 drops to a cup of epsom salts or any other mixed salts (sea salt, himalayan salt, bicarb soda).

Hair Treatment

Add 8-10 drops of essential oil per shampoo for a therapeutic treatment. Add only 2 drops per shampoo or rinse to simply perfume the hair.

Massage

Massage with the use of essential oils is deeply relaxing, invigorating and improves your well being. Add 10-15 drops of essential oil to about 30ml (~2 tablespoons) of carrier oil for a full body massage. You can use a single essential oil or mix two or more together, to suit your condition. It's recommended to blend massage oils at half strength for children, pregnant women and the elderly. And only 2-4 drops in 20ml for babies.

Do NOT massage if:

- The person is suffering cancer, a serious heart complaint, epilepsy, a fever or an acute infection.
- He/she has just eaten a big meal.
- You're tired and tense. Massage is a gift you make to another person and to do that you need to be in top spirits and full of energy.
- The area you are intending to massage covers varicose veins or a deep vein thrombosis. This would be potentially very dangerous.
- You are intending an intimate massage but have not checked that the diluted oils you are using are safe for the genital area.

5

Culinary Herbs and Aromatic Plants

Consumption Pattern

Americans are consuming ever increasing amounts of fresh, frozen, processed and dried culinary herbs and spices, and this trend appears to be here to stay. The same trend is true for specially fruits and vegetables (Fielding 1988). Factors accounting for increased consumption include interest in new foods and tastes, availability of more fresh herbs, advertising and promotion to food services and institutional food chains, and expanding ethnic populations demanding foods and flavourings of their homeland.

Who will be supplying these herbs and are there opportunities for commercial growers in the production of dried and fresh herbs? Much of the dried herbs produced domestically are produced in California by well-established food companies that own and operate drying and processing facilities and already package, transport and market these products. Growth in this area will continue as long as increased demand and consumption of these herbs continue. Additional growth is likely to occur in the contract growing of herbs by private growers to supply raw product to drying and processing companies.

Some companies both grow and process their own product, but most contract practically all of their raw material needs to growers. Processing companies that are now procuring most of their dried herbs abroad may be more willing to obtain some of the same materials domestically Growth in the dried herbal market will also probably occur with an increasing number of spice companies and natural food wholesalers and retailers catering to specialized markets (*e.g.* organic or pesticide free-herbs and spices, specially restaurants) and which unlike many of the larger food and flavor houses produce specialized herb and other food products.

There has been tremendous growth in the fresh herb market as evidenced by the increased variety of herbs available both in the larger supermarkets, the smaller grocery outlets, farmers markets, and roadside farm markets (Simon 1986, Simon et al. 1989, Simon and Clavio 1989, Simon and Grant 1987a, b). Herbs that were only available dried, or fresh during a few months of the year, are now being marketed fresh cut and more recently, as live potted herbs in the produce sections of supermarkets all year long. Recognizing this growth, the USDA (1989c) now lists weekly prices for herbs sold in nineteen major wholesale markets plus the type of container, package, and weight count, and quality of each unit in the National Market News Report. The herbs now covered include anise, arrugala, basil, borage, chervil, chives, cilantro, dill ginger root horseradish, parsley lemongrass, marjoram, mint, oregano, rosemary savoury, sage, sorrel tarragon, thyme, and watercress.

The successful introduction of new culinary herbs into commercial production requires a purposeful strategy and a solid information base. Unfortunately culinary herbs and essential oil crops have been studied little compared to all other food and fiber crops (Craker et al. 1986). While several extension guides have been published on herbs, few guides based on research are available in this country In a bibliographic review of the scientific literature from 1971-1980 (Simon et al. 1984), over 10,000 authors are listed with published research articles on the major economical herbs of the temperate zone involving horticultural research (38%), botany (17%), chemistry (14%), pharmacology (11%), ecology and germplasm (8%). With the exception of peppermint and spearmint most all crop research on herbs was conducted outside of the United States. Although the United States is the principal world producer of mint oils, only 10% of all published mint research from 1971-1980 originated in the U.S. (Craker et al. 1986).

Promising Herbs

There is a wide range of culinary herbs that can be grown in the continental United States and which may offer potential for production. Many herbs are already being commercially produced, albeit in small quantities, often in relatively small farms. Those herbs which show promise for the fresh market are listed. Three examples of culinary herbs with great promise.

Coriander (Coriandrum sativum L): This annual herb native to the eastern Mediterranean and southern European region has long been prized for its spicy aromatic seeds which are used either whole or in ground form as a main ingredient in curry and other food and flavor products.

Coriander can be grown as a seed crop in the U.S. using existing mechanization and production technology but seeds ripen unevenly and the mature seeds shatter from the plant. Present demand for coriander seed is met by existing foreign suppliers and although the yield of coriander seed is moderately high (1,100-1,700 kg/ha) and the cost of production relatively low, there is still a question whether domestic commercial opportunities exist for coriander seed production as a spice. The profit margin of coriander would at best match the more traditional cash crops (maize, soybeans). Postharvest costs of cleaning the seed, handling and shipping can be major factors determining the profitability of this seed spice crop. Coriander seed is also processed via steam distillation for the extractable essential oil of which d-linalool is the major constituent. The oil is then used in the food and perfume industries.

Recently, American consumers have witnessed the introduction of coriander leaves into the marketplace. This fresh product, known as cilantro, is marketed in bunches as a leafy green spice. While new to American cuisine, cilantro has long been a popular herb in Oriental, Middle Eastern and Latin American cooking.

Coriander, a cool season crop, is easy to grow as a culinary herb and is most suited to fertile loam soils. The plant is direct seeded with a seed drill at rates of 13-18 kg/ha in very early spring. In Florida and New Jersey, coriander is often planted weekly At harvest, the whole plant is manually cut and bunched in the field. One of the major problems in producing cilantro is premature flowering. Bolting becomes acute as the days get hotter and longer. A number of seed companies now offer slow-to-bolt (long-standing) cultivars. There are significant differences among coriander cultivars regarding the response to premature flowering, and while some are less susceptible, none are totally unresponsive to high temperatures and long days (Simon et al. 1989). Thus, cilantro is planted as a spring, early summer, or fall crop.

The aroma of cilantro is also due to an extractable essential oil, although its composition is distinctly different from the seed oil. Quality of fresh cilantro is based upon strong green colour, strong aroma, and visual appearance (*i.e.* freedom from insects, discolorations).

To the grower, yield and the cultivars' resistance to bolt are additional factors to be considered. Cilantro has a relatively short shelf life and requires refrigeration. Cilantro can be kept for 3 to 4 weeks at 0°C, but only 2 to 3 weeks when stored at 5°C (Cantwell 1989). Cilantro is routinely iced after harvest. Cilantro is expected to become increasingly important in the U.S. due in part to the expanding Hispanic and Arabic populations, its unique flavor, and increasing familiarity in the marketplace.

Sweet Fennel (Foeniculum vulgare Mill.): Sweet fennel has highly aromatic leaves and attractive green fern-like foliage but is best known for its seeds which are sold commercially as a spice. Limited domestic production for seed has occurred periodically, but demand is relatively small and adequate supplies can be obtained abroad. Successful domestic production of fennel for the seed as a spice will probably not occur unless the yields can be significantly improved, supplies from abroad become limited, or a government programme makes the production of new crops more beneficial.

Fennel is a perennial but grown as an annual. It is most suited to well drained light loam soil. The plant is easy to grow and is typically direct seeded (3.5-5 kg/ha seed/acre) into the ground in beds with a specialized planter adapted to small-seeded crops. Plant spacing varies significantly with rows in the beds 60-1 00 cm apart and with final plant stands of 10-3 cm apart in the row. Fennel is a cool-season crop, seeds are sown in early spring and germinate at temperatures >7°C (optimum soil temp. approx. 16-18°C). Yields of up to 2,240 kg/ha have been achieved, although earlier USDA estimates were only in the range of 650-900 kg/ha (Sievers 1948). The plant requires 100-115 days to mature before harvest.

The production of fennel as a culinary herb, cultivated both for its aromatic leaves and enlarged leaf stalk is on the rise. This type, called Florence fennel or finocchio fennel, is a different subspecies than regular fennel produced as a seed spice. Very popular in Europe as a specially vegetable in many culinary dishes, it is commonly consumed raw in Italy and cooked in France. Finocchio is becoming commonplace in U.S. supermarkets and consumer interest and familiarity is increasing. One of the limiting factors to increased consumption is the public's unfamiliarity with it's preparation, use and taste. Unfortunately finocchio fennel is often marketed under the misnomer of anise, another culinary herb (Pimpinella anisum L.), which has led to market and consumer confusion. Both plants contain high amounts of anethole in the essential oil, imparting the licorice-like aroma and taste. The time to harvest fennel bulbs is difficult to assess because the thickened leaf stalk continues to grow and develop until flowering takes place. Care in harvesting, grading and packing fennel must be taken to ensure a high quality fresh pack. Harvesting, cleaning, trimming and packing is done by hand. The foliage must be dark green and fresh in appearance and the stalk and bulb (enlarged base of leaf stalk) a lighter greenish-white colour. The bulb must be firm and the product free from insects and discolorations. Once harvested, fennel should be kept at 0 to 2°C (Seelig 1974). The plants are retailed individually, either wrapped in plastic or simply displayed like celery. Great variability (in growth and aroma) in commercially available cultivars of fennel and finocchio fennel exist (Simon et al. 1989) making the proper selection of cultivars for spring and fall production important. Finocchio fennel shows excellent potential for future growth in the U.S.

Oregano (Origanum spp.): Their has been a significant increase in the consumption of oregano for both the fresh and processed market (Tucker 1987). Oregano, the common name for a wide number of plant species with a characteristic aroma and flavor, is a perennial aromatic plant native to the dry calcareous soils of southern Europe, southwest Asia, and the eastern Mediterranean (Simon et al. 1984). A major problem which has limited domestic production was in the difficulty in distinguishing the many plant species and types of oregano imported and marketed as oregano.

Imported oregano was found to be derived from 16 plant genera and more than 40 plant species (Calpouzos 1954), resulting in oregano being described as a flavor and aroma rather than an individual plant. Taxonomic work by Tucker (1986, 1989) has identified the major types of commercial cultivars and cultivated taxa in this country and the major types of imported oregano. Yet much of the imported oregano arrives as a blend of plants. Selection of an individual line for domestic production, particularly for the dried leaf or as an essential oil, remains difficult. The European type of oregano comes mainly from subspecies of O. vulgare L. including ssp. hirtum (Link) Ietswaart, ssp. virens (Hoffmanns. & Link) Ietswaart and ssp. viride (Boiss.) Hayek (Tucker 1989). In contrast Mexican oregano, also called Mexican sage, is principally gathered from the small Mexican shrub, Lippia graveolens H.B.K. (Simon et al. 1984), although leaves from other species are collected.

The essential oil of European oregano is composed mainly of carvacrol and thymol. The range of each can be very wide and many chemotypes are available. The herb or the extracted oil is used in a variety of meat and sausage products, salads, stews, sauces and soups. European oregano and to a larger extent Mexican oregano, are used in flavoring Mexican foods, pizza, and barbecue sauces.

Oregano is generally transplanted to the field and grown on light, dry well-drained soils for periods of 3 to 6 years. Domestic horticultural studies on this species are limited. Although yields of more than 14 tonnes fresh herb/ha or almost 4 tonnes dried herb/ha were obtained from one commercial line by a single annual harvest (Simon et al. 1989), typically yields are much lower (1.5-3 tonnes dried herb/ha). Plants can be harvested multiple times each year (from 2 to 6) depending upon the location and end-use. Opportunities exist for dried product as well as for the fresh market. Once harvested for the fresh market oregano should be kept at 0deg.C to maximize shelf-life (Cantwell 1989).

Prospects

Production of culinary herbs for the fresh, frozen processed and dried market will likely increase. The large growth in the production of fresh culinary herbs provides opportunities to growers that either develop a market niche or cooperate closely with a broker or specialist in marketing. It is likely that the large volume herbs will continue to come from areas of intensive vegetable production such as Florida, New Jersey and California. Herbs can be expected to become fully integrated with fresh market vegetables in packaging, cooling, transport, and marketing operations. Unless governmental regulations change, there will be strong competition from Mexico, the Caribbean, and other areas to supply fresh herbs to the American public. Overproduction of specific herbs within a limited marketplace can result in significant decreases in the wholesale market prices. Greater opportunities for small producers maybe in the development of specialized markets, rather than for the wholesale trade.

Supercritical Fluid Extraction (SFE)

The supercritical fluid extraction (SFE) has been applied only recently to sample preparation on an analytical scale. This technique resembles Soxhlet extraction except that the solvent used is a supercritical fluid, substance above its critical temperature and pressure. This fluid provides a broad range of useful properties. One main advantage of using SFE is the elimination of organic solvents, thus reducing the problems of their storage and disposal in the lipidologist laboratory. Furthermore, several legislative protocols (such as the EPA Pollution Prevention Act in the USA) have focused on advocating a reduction in the use of organic solvents which could be harmful to the environment.

Besides ecological benefits, one of the most interesting properties of SFE is the high diffusion coefficients of lipids in supercritical fluids, far greater than in conventional liquid solvents. Thus, the extraction rates are enhanced and less degradation of solutes occurs. Several studies have shown that SFE is a replacement method for traditional gravimetric techniques. In addition, carbon dioxide, which is the most adopted supercritical fluid has low cost, is a nonflammable compound and devoid of oxygen, thus protecting lipid samples against any oxidative degradation.

Principles for the Analytical SFE of Lipids

The first guiding principle is the optimization of the solubility of lipids in supercritical CO_2 and the improvement of the fractionation with respect to a particular lipid species.

Important data on the solubility of vegetal oils and other lipids may be found in the detailed study of Stahl et al. (Dense gases for extraction and refining, Springer Verlag, Heidelberg, 1987). These data have led to perform oil and fat extractions above 600 bar and temperatures from 80 to 100°C. It should be emphasized that many lipid solutes have similar solubility parameters, making their separation by SFE difficult. An improvement in the separation of complex lipid mixture may be found by the integration of adsorbent compound into the extraction cell with the sample. Several compounds such as alumina, silica, Celite, Florisil or synthetic resins were proposed for the optimization of lipid differential extraction.

The improvement of lipid SFE needs also the study of the kinetics of the lipid removal from the sample matrix. It must be noticed that the fast back-diffusion of analytes in the supercritical fluid reduces the extraction time since the complete extraction step is performed in about 20 min instead of several hours. A common practice in SFE, which must be mentioned in connection with the physicochemical properties of supercritical fluids, is the use of modifiers (co-solvents). These are compounds that are added to the primary fluid to enhance extraction efficiency.

Thus, the addition of 1 to 10% of methanol or ethanol to CO_2 expands its extraction range to include more polar lipids. When the extraction was performed with supercritical carbon dioxide and 20% of ethanol, more than 80% of the phospholipids were recovered from salmon roe.

Instrumentation

Those wishing to build their own experimental equipment should read the detailed study of Hawthorne et al. (in Practical supercritical chromatography and extraction, Caude M., Thiebaut D, Eds, Harwood Academic, 1999).

ISCO Inc was one of the first companies to address the off-line SFE market. Extraction cells of 0.5 up to 15 ml are offered.

An automatic version permits automated valve operation for extraction of one or two samples and another model permits 24 samples to be automatically and sequentially extracted. ISCO has also offered a extraction system (FastFat HT) specifically made for the rapid determination of fat content in foods and agriculture products ("The Isco FastFat HT provides rapid, accurate % fat determination in a wide range of foods- a critical parameter in quality control, product labeling, and calibrating on-line process control equipment.

With FastFat HT, you get reliable results in minutes, and eliminate costly delays in production decisions").

Applied Separation offers several extraction units offering great flexibility with respect to sample size and experimental design. Extractor cell sizes can range from several ml to one liter. Off-line and in-line trapping of solutes may be realized. IECO Corporation is offering a total fat analyzer designed with a triple parallel channel system ("LECO's new TFE2000 Fat/Oil Determinator is specifically designed to determine fat/oil content and eliminate the need for hazardous chemicals used in Soxhlet, Mojonnier/Roese-Gottlieb, acid hydrolysis, and other solvent extraction methods.

Using inexpensive compressed CO2 as the solvent, the analyte is extracted from the sample and transferred to the removable ultra efficient solid phase collection traps. Precise calculations are determined by the integrated balance based on sample weight, and the before and after weights of the collection vials"). Application notes may be found on the LECO site. The use of high purity SFE-grade CO_2 is not required but impurity and moisture in industrial grade CO_2 can accumulate and may interfere with further analytical operations (gas or liquid chromatography). Thus, an on-line fluid cleanup system may be used to remove trace contaminants.

Extractions can be performed in static, dynamic or recirculating mode. During static extraction, the cell is filled with the supercritical fluid, pressurized and allowed to equilibrate. In the dynalic mode, the fluid is run continuously through the cell, and in the recirculating mode, the same fluid is repeatedly pumped in the cell before being pumped out to the collection vial.

Important Parameters

A. *Sample Matrix:* The levels of lipid content and moisture are important for the analytical process. Lipid levels may be from 1 up to 50% (w/w).

B. *Sample Preparation:* Moisture may inhibit contact between extraction fluid and sample. Thus, removal of water by freeze drying is recommended prior to SFE.

It may be also efficient to disperse the sample matrix, to grind the sample to increase the mass transfer of lipids. The choice of drying agent can be made by consulting the study of Bulford (Bulford MD et al., J chromatogr A 1993, 657, 413).

C. *Collection of the Lipid Extract:* The collection method for the resultant extract must be optimized to avoid incomplete extraction. The open vials are most frequently used, a sub-ambient cooling being useful for volatile species. It is best to use a collection vial packed with a surface area material, *i.e.* glass beads, glass wool (Snyder J. M. et al., JAOCS 1994, 71, 261) or containing some volume of a chosen solvent. The supercritical CO2 can be separated from the collected analytes without trouble or disposal problems.

Application of SFE to Lipids

Several reviews were devoted to the application of SFE for the extraction of lipids in various sample types. Those of Clifford A et al. (in Supercritical fluid technology in oil and lipid chemistry, AOCS Press, chap. 19, 1996), Eller FJ et al. (Sem Food Technol 1996, 1, 145) and King J.W. (Grasas y Aceites 2002, 53, 8) should be consulted for specific applications. Food and various agricultural products were successfully analyzed for their fat and levels by SFE. Progressively, SFE may be regarded as an alternative to organic solvent extraction methods. The extraction efficiencies of SFE and organic solvent methods (Soxhlet or liquid-liquid extractions) were frequently compared and shown to be in good agreement (Eller FJ et al., J Agric Food Chem 1998, 46, 3657). It must be noticed that gravimetric-based results can be influenced by the sample matrix, the moisture and non-lipid moieties and the extraction solvent.

Total fat determinations were done in meat products with two different extractors and compared with a standard method. The amount and the composition agreed well with results from the standard procedure. To obtain quantitative recoveries by SFE, 1 ml ethanol was added to 1 g sample in the extraction cells before extraction.

Extracting conditions largely affected the composition of dissolved lipids in freeze-dried animal samples (Tanaka Y et al., J Oleo Sci 2003, 52, 295). Thus, one group of triglycerides was extracted from salmon roe while another group remained in the matrix together with phospholipids and a large part of astaxanthin. It is now well known that the extraction yield and composition of triglycrides depend on the extracting conditions. This was verified for the extraction of oil from seaweed, tomato seeds or soyabean oil.

The extraction and separation of phospholipids from tuna fish have been described using various concentrations of methanol in supercritical CO_2 (Tanaka Y et al., J Oleo Sci 2005, 54, 569). Good recovery of DHA-rich phospholipids were claimed in a series of industrial processes. A detailed description of a reliable procedure adapted for the determination of total fat in milk- and soy-based infant formula powder has been published by LaCroix DE et al. (J AOAC Int 2003, 86, 86).

The extraction of a specific lipid moiety is generally influenced by substances which are co-extracted, all having a high solubility in supercritical CO_2.

One technique to overcome this problem is to use a sorbent to retard the analyte of interest, the interfering substances being removed first. Then, a higher CO_2 extraction density or a co-solvent is used to removed the analyte of interest from the sorbent. Thus, it was shown that NH2-bonded silica is able to retard sterols.

Analytical SFE has also been used for lipid-derived volatiles such as aroma in various food samples. For the application of SFE to the study of total fat and lipid classes in meats, the studies of Berg H et al. (J Chromatogr A 1997, 785, 345) and Chandrasekar R. (JAOAC int 2001, 84, 466) should be consulted.

Field of the Invention

The extraction of the principal significant components of herb and spice plants containing pigment, flavor, and aroma and, when present in the starting material, antioxidant, using edible, food-grade solvent.

Of especial interest is the extraction of the principal significant components of spices and herbs represented by plants of the family Solanaceae, representatively Capsicums such as paprika, red pepper, and chili, and Lycopersicon, representatively tomato, all containing carotenoid pigments; Umbelliferae, representatively celery, lovage, dill, carrot, cilantro, fennel, cumin, caraway, parsley, angelica, and anise; Compositae, representatively marigold, artemisia, and tarragon; Leguminosae, representatively fenugreek; Labiatae, representatively rosemary, thyme, sage, oregano, marjoram, mint, savoury, and basil; Zingerberacae representatively ginger, cardamom, and turmeric, Lauraceae, representatively laurel, cinnamon, cassia, and bay; Myrtaceae, representatively allspice and clove; of the genus Myristica, representatively mace and nutmeg; of the genus Piper, representatively black and white pepper; of the genus Vanilla, representatively vanilla; of the genus Allium, representatively onion and garlic; of the genus Sesamum, representatively sesame seed; Cruciferae, representatively mustard and horseradish; the defining characteristic being that it is of a spice or herb plant material from which flavor, aroma, colour, and/or antioxidant can be extracted and used to flavor and/or colour foods and beverages or otherwise employed to enhance the palatability of foods and beverages.

Background of the Invention and Prior Art

The present invention relates to a method of increasing the stability and reducing the microbial counts of both spice or herb oleoresin and the residual cake from which the oleoresin has been extracted. The process simultaneously extracts and concentrates the principal flavor, aroma, colour, and other active components, produces a concentrated and standardized food-grade extract of active components, and a standardized food-grade residual solid, with both the extract and residual solids having significantly reduced microbial counts and frequently also improved stability.

Concentrated extracts of spices and herbs are universally used for flavoring and coloring of food, beverages, and pharmaceuticals. These extracts are traditionally used where a standardized, sterile, and uniform concentrate offers the benefits of control which are inherently difficult to obtain from raw spice or herb, or where the bulk of the raw material is not needed or undesirable.

Ground spice and herb solids are universally used for flavoring, coloring, and imparting otherwise favourable characteristics to food and beverages where the bulk, functional characteristics, and appearance of the food or beverage is important.

Dried spices and herbs, most often in their ground form, are used in the preparation of food and beverages to add flavor, aroma, colour, and preservative properties that make the food more palatable and appealing. The dried spices, ground or unground, are usually added to the food or beverage during preparation at such a point in the preparation that time is allowed for the principal components of interest to be extracted into the food or beverage to impart the desired combination of attributes to the food or beverage. Further, as spices and herbs are notoriously known to have inconsistent levels of the flavor, aroma, colour, or antioxidants, it is commonly required that spices of varying levels of the principal components of interest be blended to make a final product that is consistent with regard to the principal components of interest to achieve predictable and repeatable performance with respect to the flavor, aroma, colour, or antioxidant release into the food or beverage system in which they are used. This is a costly and time consuming process.

As suggested in prior art, much of the flavor, aroma, and/or colour often is not effectively transferred to the food or beverage. U.S. Pat. No. 2,507,084 overcomes this obstacle of under-utilization of the principal components of interest by first extracting the principal components of interest and subsequently coating the spent spice from the extraction process with a portion of the extract originally removed, thereby extending the useful amount of flavor and aroma that can be derived from a given quantity of spice. It is also disclosed that this process derives value from the exhausted spice solids, from which the flavor, aroma, colour, or antioxidants have been removed, which would otherwise be a waste product. This is a complicated and costly process for recovery of the maximum value of the spice and its principal components of interest.

Traditional extraction processes for the manufacture of concentrated extracts (concentrated several fold as compared with the raw material) involve not only the use of various non-edible solvent systems, but also a large proportion of solvent in relation to the compounds of interest. Many require the use of petroleum distillates, chlorinated solvents, or highly flammable solvents which must be eliminated almost completely from the finished products to make them safe for consumption. These systems require expensive distillation equipment and special precautions must be taken to ensure worker safety and to limit environmental impact.

The intensive processing required often destroys, modifies, or loses some of the more unstable compounds, delicate aromas, flavours, or pigments. More significantly, the last traces of undesirable non-edible solvents are very difficult to separate from the concentrated extract. The residual solid must necessarily contain the same residual non-edible solvents, which are removed only with difficulty. Such residual solvents limit the potential use of the residual solid for human consumption, and are potential environmental contaminants.

Other concentration techniques rely on high pressure equipment to obtain good solvating properties from gases, *e.g.*, liquid or supercritical CO_2 (U.S. Pat. No. 4,490,398). High pressure liquefied or supercritical gas extraction requires expensive equipment and has limited solvating abilities for some compounds requiring the addition of cosolvents, or solvents such as propane and butane, which are also difficult to control and may be environmentally sensitive or undesirable in a finished product.

Following extraction and desolventization, the concentrated extract is often standardized with edible solvents and emulsifiers to provide a concentrate with reproducible levels of the active or principal compounds of interest to the user. In an effort to overcome the shortcomings and risks associated with the above-mentioned processes, extraction has been carried out using edible solvents such as vegetable oils or lard. Typical extraction procedures are disclosed in U.S. Pat. Nos. 3,732,111; 2,571,867; and 2,571,948. These methods require a relatively large volume of solvent in relation to the compounds of interest and result in a dilute extract which is limited in its application and which has few of the advantages of the concentrates which can be produced using volatile solvents.

U.S. Pat. No. 4,681,769 discloses a method for simultaneously extracting and concentrating in a series of high pressure countercurrent mechanical presses using relatively small amounts of vegetable oil as the solvent in an attempt to overcome the problem of dilution inherent in earlier processes. This method suffers from severe limitations in temperature and pressure ranges in an attempt to avoid unacceptable oxidative damage, colour loss, yield losses, and flavor changes with the final result being that contact times must be unduly extended for up to 16-24 hours, adding greatly to the cost of the process.

Extraction cycle times are unduly long for a given size pressing operation, and the process does not provide for a controlled degree of browning or for sterilization of the extract or of the residual solid. It is also limited to temperatures of less than 100° F. to avoid alleged undue oxidation and thus it does not allow for the use of edible solvents which have a melting point of more than 100° F. or which are highly viscous at temperatures of less than 100° F. Maximum pressures of up to about 500 PSI (cone pressure) are claimed and this severely limits the efficiency and throughput rate for a given size pressing operation, as shown by the disclosure of this patent.

Traditional methods for the sterilization of ground spices and herbs involve the use of extremely toxic substances such as ethylene oxide or methyl bromide, irradiation, or steam and moisture treatment to reduce plate counts to less than 100,000. Chemical sterilization and irradiation of spices and herbs are disagreeable to the consumer because of the perceived risk of residual chemicals and/or radiation remaining in the plant matter and, as a result, several processes using added moisture, such as water or steam, at elevated pressures have been developed as alternatives.

Typical sterilization procedures are disclosed in U.S. Pat. Nos. 4,210,678, 4,790,995, and 4,910,027. All sterilization processes are inherently costly in that they require a separate processing step or steps to accomplish the sterilization, and also present the possibility of further degrading the more unstable components. Addition of moisture or water vapour, as disclosed in U.S. Pat. Nos. 4,210,678 and 4,910,027, prior to or during the heating and sterilization process results in a cooked aroma not typical of the fresh, dehydrated spice or herb and also results in steam distillation and loss of some of the volatile flavor and aroma constituents.

U.S. Pat. Nos. 4,790,995 and 4,910,027 require the addition of a coating of animal protein to protect the spice from the loss of volatile aroma compounds during the sterilization process with water vapour. U.S. Pat. No. 4,210,678 requires bringing the moisture of the spice to above 8-14%, in some cases up to 16-20%, and holding the spice for an extended period of time prior to sterilization to equilibrate the moisture. This additional step is costly and time consuming. In the case of Capsicums, severe browning and off aromas and flavours are developed in the presence of moistures in excess of 10% at elevated temperatures above 180° F.

Traditional methods for controlling the brownness or degree of caramelization of Capsicum solid to enhance its visual appearance involve the use of elevated temperatures and the addition of vegetable or animal fats or oils to bring up the surface colour and luster of the ground spice. This requires a separate and costly processing operation.

Above all, there is the unsolved problem of obtaining satisfactory yields, quality, and throughput rates of acceptable extract having an acceptable content of active principle in the edible solvent without undesirable oxidative damage to, and reduced stability of, the principal compounds of interest, while at the same time providing for simultaneous sterilization of both the herb or spice solid and extract. Obviously, existing prior art procedures leave much to be desired, and it is a primary objective of the present invention to provide a procedure for the production of sterilized spice and herb products having enhanced stability and which otherwise obviates the shortcomings of the prior art.

Objects of the Invention

Accordingly, it is an object of the present invention to provide a process for simultaneously and rapidly extracting and concentrating the principal components of herb and spice solids, at temperatures of at least 130° F., preferably 130 to about 450° F., in a process which is completely free of petroleum, chlorinated or highly flammable solvent, does not require high pressure gas handling equipment, does not require distillation for solvent removal, uses only food-grade edible solvents which are typically used in the trade to standardize the resulting extract to a desired concentration, and provides a product of reduced bacterial count which is free of adulterants and impurities.

Another object of this invention is to prepare such a concentrated extract by a process which is simple, environmentally friendly, and economical. A further object of this invention is to prepare a residual solid or press cake which is edible, free of residual petroleum distillates, chlorinated solvent, or other adulterants, which is standardized with respect to the principal components of commercial interest, which has a predictable and controlled degree of brownness or caramelization, and which has a controlled level of water activity with its attendant increased resistance to oxidative deterioration of carotenoid pigments and colour loss where this is otherwise a problem.

Still another object of this invention is to prepare a residual solid or press cake that is standardized with respect to the principal components of interest, that when reground rapidly and efficiently imparts flavor, aroma, colour, and/or other principal components of interest to a food or beverage in which it is used, and all without the use of undesirable non-edible solvents that are inherently difficult to remove from spice solids remaining after extraction prior to standardization with respect to the principal components of interest, are environmentally unfriendly, and are perceived by the consumer as being undesirable in the preparation of a food ingredient. A still further object of this invention is to provide such a process wherein antioxidants can be added to the edible solvent system so as to protect the concentrated extract and the residual solids against oxidative degradation of the principal components of interest, *i.e.,* flavor, aroma, and colour, which are extracted from the raw plant material or left in the residual solids.

Still a further object of this invention is to prepare an edible extract and an edible residual solid with reduced microbial activity by a process wherein the moisture of the herb or spice is kept below 10%, preferably below 8%, thereby avoiding the loss of volatile flavor and aroma constituents and avoiding the development of uncontrolled browning and off flavor development at temperatures in excess of 130° F. which are necessary to effect high extraction efficiencies, reduction in microbial activity, and improved stability of the carotenoid pigments in both the extract and in the residual solids.

Yet a further object of this invention is to prepare an extract with increased resistance to oxidative degradation of pigments and consequent colour loss. Still an additional object is to provide a process wherein and whereby the residual spice or herb plant solids have their tissue ruptured so as to produce quick release of the flavor, aroma, colour, or antioxidant component therein and thereof when in use in a food or beverage thereby greatly enhancing its use effectiveness, and whereby all products of the process may be conveniently standardized with respect to the principal flavor, aroma, colour, or antioxidant component of interest.

Other objects will be apparent to one skilled in the art to which this invention pertains and still others will become apparent hereinafter as the description proceeds.

Brief Description of the Drawings

The process of the present invention, including the several process steps involved in the simultaneous extraction and concentration of herb or spice solids, *e.g.*, Capsicum solids, to produce the desired extract and sterilized residual solid, both of which have increased resistance to oxidative degradation, and which can be readily standardized to desired levels of the principal components of interest. Although the process illustrated comprises three extraction stages, the number of stages can be decreased to two or increased to more than three to effect the desired relative principal component concentration in the extract and in the residual solid.

The invention, then, inter alia, comprises the following, alone or in combination:

A continuous multistage mixing, high pressure pressing, and countercurrent extraction process for the production of a concentrated edible extract and quick-release edible residual solids, both of reduced bacterial content, and both of which contain herb or spice pigment, flavor, and aroma and, when present in the starting material, antioxidant, from herb or spice plant solids, comprising the following steps: subjecting said herb or spice solids to a countercurrent extraction process involving a plurality of mixing and pressing stages, including first and last mixing stages and first and last pressing stages, together with up to about fifty percent by weight of an edible solvent, to produce an extract and residual solids, continuously returning the extract from each pressing stage to the previous mixing stage, and finally separating the extract from the first pressing stage and separating the residual solids from the last pressing stage, all pressing stages being carried out at a temperature of at least 130° F.; such a process wherein the temperature is 130 to about 450° F.; such a process wherein the solids are subjected to internal pressures in the press stages of at least 6,000 pounds per square inch; such a process wherein the weight of the edible solvent is 5% to about 20% by weight of the solids; such a process wherein the moisture content of the starting solids is less than 10% by weight, and wherein bacterial count reduction is effected at this low moisture content, thereby avoiding undesirable loss of volatile flavor and aroma constituents and avoiding the development of cooked, off flavours and aromas which occur at higher moisture contents; such a process wherein the solids extracted in the process are selected from the group consisting of chipotle, turmeric, black pepper, onion, rosemary, and oregano; such a process wherein the edible solvent is selected from the group consisting of soybean oil, corn oil, cottonseed oil, rapeseed oil, peanut oil, mono-, di-, or triglycerides, lecithin, edible essential oils, sesame oil, edible alcohols, hydrogenated or partially hydrogenated fats or oils, polyoxyethylene sorbitan esters, limonene, edible animal fats or oils, mixtures thereof, and edible derivatives thereof; such a process wherein fine particulate solids are filtered or centrifuged from the extract and alternatively discarded, returned to a mixing or pressing stage of the process, or incorporated in the final residual solids; such a process which includes the steps of hydrating the final extract to add water to the extent of 5% to 200% by weight of the gums and fine particulate solids therein and filtering or centrifuging to remove said gums and solids; such a process including the step of returning the separated hydrated gums and solids to the final residual solids; such a process including the step of rehydrating the final residual solids with water to a water activity greater than 0.3 AW for stabilization thereof; such a process wherein the solids are rehydrated to a water activity of about 0.4 to 0.6 AW; such a process wherein an effective stabilizing amount of an edible antioxidant or chelator is included in the edible solvent; such a process wherein the antioxidant comprises an antioxidant selected from the group consisting of lecithin, ascorbic acid, citric acid, tocopherol, ethoxyquin, BHA, BHT, TBHQ, tea catechins, sesame, and the antioxidant activity from an herb of the Labiatae family; such a process wherein the antioxidant comprises a naturally-occurring antioxidant from an herb of the family Labiatae or powdered ascorbic acid; such a process wherein the antioxidant comprises the antioxidant activity from an herb selected from the group consisting of rosemary, thyme, and sage; such a process wherein the temperature is greater than 180° F.; and such a process wherein the temperature is between about 180° F. and 235° F.

Moreover, an extract of herb or spice plant solids produced by the process having a high level of principal flavor, aroma, colour, or antioxidant components of interest and a low bacterial count due to the high temperature employed in its production and due to the low water content not greater than 10% in the starting solids; and an extract of herb or spice plant solids having a low bacterial count due to the high temperature employed in its production and having improved stability produced according to the process due to edible antioxidant therein; and residual spice or herb plant solid having a low bacterial count due to the high temperature employed in its production and having its tissue ruptured so as to produce quick release of the principal flavor, aroma, colour, or antioxidant component therein, and which is standardized with respect to the principal flavor, aroma, colour, or antioxidant component of interest, produced by the process of the present invention.

The Present Invention

In General

Raw Capsicum or other spice or herb solids, either ground (usually to pass US 40 mesh, and preferably to pass at least US 20 mesh) or unground if coarse particles are desired in the residual solid or cake, *e.g.*, spice solids having a moisture range of about 0.5% to 16% by weight, preferably 0.5 to 12%, and most preferably 1.5% to 10% by weight (ASTA method 2.0), are subjected to a mixing stage, preferably high shear, and in at least one stage an edible solvent is thoroughly dispersed throughout the raw plant material solids.

Typical spice starting plant materials include, for example but without limitation, those of the genus Capsicum including the dried ripe fruits of Capsicum frutescens L. (chilies), Capsicum annum L. (Spanish peppers), Capsicum annum L. var. longum Sendt, its hybrid Louisiana Sport Pepper, and Capsicum chinense (Scotch Bonnet or habenero), all by way of example and not by way of limitation.

Other spices and herbs of interest include those of the family Solanaceae, as stated representatively Capsicums such as paprika, red pepper, and chili, and Lycopersicon, representatively tomato, all containing carotenoid pigments; Umbelliferae, representatively celery, lovage, dill, carrot, cilantro, fennel, cumin, caraway, parsley, angelica, and anise; Compositae, representatively marigold, artemisia, and tarragon; Leguminosae, representatively fenugreek; Labiatae, representatively rosemary, thyme, sage, oregano, marjoram, mint, savoury, and basil; Zingerberacae, representatively ginger, cardamom, and turmeric, Lauraceae, representatively laurel, cinnamon, cassia, and bay; Myrtaceae, representatively allspice and clove; of the genus Myristica, representatively mace and nutmeg; of the genus Piper, representatively black and white pepper; of the genus Vanilla, representatively vanilla; of the genus Allium, representatively onion and garlic; of the genus Sesamum, representatively sesame seed; Cruciferae, representatively mustard and horseradish; the defining characteristic being that it is a spice or herb plant material from which flavor, aroma, colour, and/or antioxidant can be extracted and used to flavor and/or colour foods and beverages or otherwise employed to enhance the palatability of foods and beverages.

The comminuted or uncomminuted plant material is subjected to a plurality of mechanical pressing stages, whereby a concentrated extract of principal components is obtained and a final utilizable and preferably standardized residual solid is produced. The selected edible solvent is introduced into the residual solid at a mixing stage at some point prior to the last pressing stage. The edible solvent, now containing extract, is cycled back to the previous stage, thus always supplying a solvent extract with increasing principal component concentration to the previous mixing and pressing stages. As the extract/edible solvent is passed through each stage countercurrent to the solids flow, a portion of the edible solvent is squeezed or pressed out, thereby extracting a portion of the principal components of interest. As the edible solvent/extract passes countercurrent to the solids, the extracted principal components are progressively concentrated in the extract in a continuous process and the residual principal components end up in the final residual solids known as the cake.

By varying the pressure, temperature, spice solids feed rate, solvent addition rate, and the number of mixing and pressing stages, the concentration of the principal components can be controlled in both the extract and the residual solid. As will be apparent to one skilled in the art, variations in the process of the present invention can be employed to produce variations in result, the most advantageous of which are the production of both plant material extract of standardized marketable potency and edible residual solid plant material also characterized by standardized marketable potency, and with the edible residual solids also characterized by rapid release of the principal components of interest. Moreover, in the process of the present invention, the spice or herb plant solids have their tissue ruptured so as to provide quick release of the flavor, aroma, colour, or antioxidant component therein when in actual use, *i.e.*, when incorporated into foods or beverages.

For example, using 200 ASTA paprika starting material of about 5% moisture, a 20% soy oil addition, and leaving a residual cake extractable yield of 9.8% by weight of the starting plant solids material, gives an extract with a colour value of 850 ASTA and a residual cake colour value of approximately 50 ASTA. Contrastingly, using a 10% soy oil addition (instead of 20%) yields a cake having about 65 ASTA colour value and, by increasing the residual cake extractable yield to 12.5% by weight of starting plant solids material, the colour value of the residual cake rises to about 100 ASTA and that of the extract to about 1400 ASTA. The lowest colour extract for paprika normally traded is 1,000 ASTA.

Although less than 20% edible oil addition is highly desirable and can be used in many cases, with some edible solvent systems wherein the principal compounds of interest have a limited solubility, or when a more dilute extract and/or lower concentration of principal compounds is desired in the residual solids, more than 20% by weight of edible solvent addition will be required inasmuch as a suitable concentration of principal compounds in the finished extract and in residual solid can in some cases be produced only by the employment of the higher dilution.

Due to the successive treatments of high pressure and pressure relief, with pressures ranging from 6,000 to 30,000 PSI in the pressing stages of the operation, in the presence of added edible solvent, *e.g.*, vegetable oil, and due to frictional heat generated in these high pressure zones, both the residual solid and the extract exiting the process surprisingly have a significantly reduced microbial load over that of the starting material even at moisture levels significantly lower than those indicated by the prior art and, also surprisingly, exhibit increased resistance to oxidative degradation of the carotenoid pigments which are responsible for the characteristic red-yellow colour of various spices, *e.g.*, Capsicums.

The extract from the first or any selected pressing stage may be centrifuged or filtered to provide the finished extract free of particulate solids. Preferably, the fine particulate solids and gums in the extract may be hydrated to about 5% to 200% by weight of the gums and solids prior to centrifugation or filtration to give a crystal clear extract. If water is not used to hydrate the solids and gums, the fine particulate solids from the extract may conveniently be combined with the final residual solids, recycled back into mixing and pressing stages of the process, or alternatively discarded. If water is used to hydrate the solids and gums, it is preferred that the solids and gums be added back to the final residual press solids or discarded.

The edible solvent employed according to the process of the present invention, as illustrated by the following Examples, may be any edible solvent and especially those selected from the group consisting of soybean oil, corn oil, cottonseed oil, rapeseed oil, sesame oil, peanut oil, mon-, di-, and triglycerides, lecithin, essential oils of spices, herbs, or other plants, edible alcohols, propylene glycol, glycerine, hydrogenated or partially hydrogenated fats or oils, limonene, polyoxyethylene sorbitan esters, or any other edible vegetable or animal fat or oil, or mixture thereof, or edible derivatives thereof, the essential aspects of the solvent being that it serves as an extraction aid in which the principal components of the material being extracted are soluble and that it be edible.

The edible solvent, according to the present invention, is combined with the raw material solids to be processed in a proportion of about 5% to about 50% by weight, and frequently amounts as low as 5 to 20% by weight are possible, based on the weight of the starting raw material solids to be extracted. The lower percentages frequently produce a more acceptable and marketable concentration of principal components of interest in both the extract and the residual solids. The temperature to be employed during the processing and especially in the pressing stages of the process of the invention may be varied widely, but the process is generally carried out at a temperature below about 450° F., and between about 130° F. and 325° F., most preferably above 180° F. and especially between about 180° F. and 235° F.

Temperatures in excess of 130° F. are advantageously employed to achieve acceptable yields and increased throughput rates as compared to the prior art. Higher temperatures are employed to control an increased degree of browning and, most importantly, to reduce the microbial load of both the solids and the extract while at the same time imparting increased resistance to oxidative degradation of the carotenoid pigments of Capsicums in both the extract and the residual solids. Thus, when it is desired that the residual solids from the process have a desirable darkened, caramelized appearance and/ or flavor, a reduced microbial load, and increased resistance to oxidation, this is readily attained by increasing the temperature of the solids and the extract during the process, especially during the pressing stages thereof.

When an antioxidant or chelator is introduced into the process for protection of the spice or herb being processed, this is preferably another plant material or an extract thereof, preferably of the Labiatae family, such as rosemary, thyme, or sage, which is known for its protective antioxidant activity (U.S. Pat. No. 5,209,870), or sesame, or tea catechins, but may alternatively be a suitable edible and preferably an approved food grade additive such as ethoxyquin, BHA, BHT, TBHQ, tocopherol, Vitamin C (*e.g.*, as in U.S. Pat. Nos. 5,290,481, 5,296,249, or 5,314,686), citric acid, EDTA, or the like. The process of the present invention is particularly adaptable to the extraction of any spice or herb plant material solids containing carotenoid pigments or other components which provide colour and/or flavor, pungency, aroma, or antioxidant activity, such as present in rosemary, thyme, sage of the Labiatae genus, to a food with which combined.

Detailed Description of the Invention

The following examples are given to illustrate the present invention but are not to be construed as limiting.

Example 1: Paprika Extraction: Dehydrated paprika (5.5% moisture) is ground in a hammer mill and the resulting ground paprika (95% passing US 40 mesh) is admixed with about 10% by weight of soy bean oil and processed in a countercurrent extraction system involving three (3) pressing stages, each using an Egon Keller Model KEK-100 Screw Press, with the extracts from the second and third stages being returned to the preceding mixing stage before being removed from the process at the end of the first press stage. A high shear, high speed pin mixer or equivalent is used to mix the soy oil or extracts from the second and third press stages into the ground spice or residual solid from the preceding stage. This recycling is continuous. The raw material paprika solids are continuously fed at a rate of about 240 lbs. per hour with a total contact time in each mixing stage of about 15-60 seconds. The residence time in each press is 5-60 seconds. The pressing stages are operated at about 10,000 PSI internal pressure and about 200 degrees Fahrenheit, which is maintained by cooling with water through the bore of the press shafts.

The starting colour value of the ground paprika solids is 200 ASTA. The principal components extracted and standardized in both the extract and the residual solid are the carotenoid pigments.

The resulting final soy-paprika extract has a colour value of about 1,375 ASTA and the reground paprika residual solid from the final (3rd) press stage has a colour value of about 85 ASTA.

Example 1A: Variation: By varying the percentage of edible solvent employed from about 5% to 20%, the pressure from about 6,000 to 30,000 PSI, the number of countercurrent mixing and pressing stages from 2 to 5, with return of the extract from each press stage to the preceding mix stage before final removal from the process in the first press stage, varying the temperature from about 130° F. to 280° F., and removing the seed from the paprika solids prior to grinding, the resulting extract ranges in colour value from about 2,700 ASTA to about 800 ASTA and the residual solids range in colour value from 180 ASTA to 35 ASTA.

By regrinding the residual solids (from the final stage) just as is done with fresh, dehydrated paprika, a product in every way comparable to commercially available ground paprika solids is produced. After filtering or centrifuging off the fine particulate solids, the extract can be directly substituted for commercially available paprika oleoresin in every respect. By varying the pressing temperature of the process from about 130° F. to 325° F., the hue of the reground residual solid is varied from slightly browned to a dark chocolate brown, demonstrating that the degree of brownness can be controlled by the pressing temperature employed.

The degree of "brownness" is measured using a Hunter Labscan Spectrocolorimeter with 0 degree illumination, 45 degree circumferential viewing, illuminant D65, 10 degree observer, Ceilab coordinate system. The hue of the paprika powder is measured by placing the powder in a 2.5-inch diameter cuvette, shaking gently to ensure even coverage, and measuring through the bottom of the cuvette. The results of the varied operating temperatures of the process are shown in Table. The designation L is indicative of the "lightness" of the sample with the higher numbers being lighter or less browned, and the lower numbers being darker or more browned.

Processing Temperature	*Visual Appearance*	*L* Values*
130° F.	Red	40.18
150° F.	Tan-Red	37.25
200° F.	Light Brown Red	33.22
280° F.	Dark Brown Red	29.16
325° F.	Chocolate Red	22.85

The data clearly demonstrate that the degree of browning can be controlled by varying the press temperature at which the process is conducted. This broadens the applications or uses of the residual solid to include a base for toasted chili powder and as a replacement for browned, caramelized paprika. The residual solid can be substituted for ground paprika or chili powder in many common applications and a separate processing step for browning to a desired degree is not required.

The starting ground paprika solids have an aerobic plate count (Analysis run according to Bacterial Analytical Manual By AOAC, 8th edition, 1995, and ISO-GRID Methods Manual, 3rd edition, 1989) of about 14,000,000. The residual solids exiting the extraction system have a count of about 2,000 to 200,000, with the lower count being achieved at the higher temperatures. This is a significant reduction and makes the residual solids per se suitable for any application where treatment with ethylene oxide or irradiation would normally be required.

Example 1B: Antioxidant Addition: The foregoing example is repeated with all materials and conditions being the same, except that the soybean oil edible solvent is supplemented with an antioxidant blend at a concentration of 3% by weight of the original ground paprika solids. The blend consists of about 29% lecithin, 20% powdered ascorbic acid, 5% citric acid, 15% tocopherol, and 1% rosemary extract (in accordance with Chang and Wu U.S. Pat. No. 5,077,069).

The stability of (1) the resulting extract and (2) the residual solids is compared in each case with an untreated control. In such evaluation, the paprika extracts are plated on flour salt to an extent of 2.4% by weight with a mortar and pestle. Two-gram samples are weighed into 13×100 mm test tubes. The test tubes are stored in a thermostatically-controlled oven at 65° C. Samples are withdrawn periodically, extracted with acetone, and the colour at 460 nm of a standard (%) dilution in acetone is determined spectrophotometrically. In the evaluation of the residual solids, two-gram samples of the reground residual solid are substituted for the flour salt dispersions.

The procedure for the "standard dilution" is as follows: The initial colour of the dispersion is determined by pouring two grams of the original dispersion into a 100-ml flask. Acetone is added up to the 100-ml level. The flask is inverted several times. The flour salt is allowed to settle for five minutes. Then three ml of the dilution is pipetted into a 25-ml flask and diluted up to the 25-ml level. The absorbance is read at 460 nm. The 460 nm colour is determined by the formula: ##EQU1## to translate to ASTA colour, multiply the 460 nm colour by 820. The colour is plotted against time and the time for 1/3 of the starting colour to fade is reported as the 2/3 life.

This is a highly-reproducible measurement, which is sufficiently accurate to evaluate the effectiveness of the antioxidants and will assist the practitioner to optimize formulations for specific uses. The final extract from the first press stage of the unprotected or unstabilized process has a colour value of about 1375 ASTA and a 2/3 life of 6.5 hours as compared to a colour value of about 1600 ASTA and a 2/3 life of 63 hours for the extract from the protected material.

The colour value of the unprotected or unstabilized residual solids is about 85 ASTA with a 2/3 life of 54 hours, compared to the protected solids which have a colour value of about 95 ASTA and a 2/3 life of 155 hours. This clearly demonstrates that inclusion of antioxidants can improve not only the colour yields from the extraction process but also at the same time improve the colour stability of both the extract and the residual solids.

Other suitable antioxidants (*e.g.*, lecithin, ethoxyquin, butylated hydroxy anisole (BHA), butylated hydroxy toluene (BHT), tertiary butyl hydroxy quinone (TBHQ), sesame, tea catechins, and Labiatae herb antioxidant activity, finely-divided ascorbic acid, tocopherol, citric acid) can be substituted in whole or in part for the specific antioxidant mixture employed with similar desirable colour-protective results, preferably a naturally-occurring antioxidant from an herb of Labiatae family, *e.g.*, rosemary, sage, or thyme, or powdered ascorbic acid.

Example 2: Effect of Varying Operating Temperatures: Dehydrated paprika solids (2.5% moisture) were ground in a hammer mill and the resulting ground paprika (95% passing US 40 mesh) was processed with about 15% by weight of soy bean oil in a countercurrent extraction system as in Example 1 involving two (2) pressing stages, with extracts from the second press stage being returned to the preceding (first) mix stage before being removed from the process at the first press stage. Upon exiting the first press stage, distilled water was metered continuously into the crude extract at a rate of 75% by weight of the gums and solids by means of an inline static mixer. The weight of the gums and fine particulate solids in the extract was determined by diluting one gram of the crude extract in nine grams of acetone. The mixture was spun down for three minutes at 2000 G's in a laboratory centrifuge.

The solids separated were air dried and the weight of the gums and solids was calculated as a percentage of the weight of the starting extract. The hydrated gums and solids removed from the extract were continuously returned to the final residual press solids via a high shear, continuous pin mixer installed immediately following a water-jacketed cooling screw which received the residual solids from the second press stage. Prior to hydration and centrifugation, the extract contained approximately 10% by weight of gums and fine particulate solids as determined by the above-described method. Following hydration and centrifugation the gums and particulate solids amounted to no more than 1% by weight of the extract and the extract was a crystal clear solution, free of any suspended insoluble materials.

The colour value of the starting ground paprika was about 150 ASTA. The pressing stages were operated at about 20,000 to 30,000 PSI. The extraction process was started with the presses operating at about 80° F. as measured by the temperature of the cake exiting the presses. The temperature of the presses was controlled by the rate of flow of cooling water through the bore of the press shafts and the screen cages to keep the operating temperatures in the range of 80° to 180° F. Over the time of the extraction run, the operating temperatures of the presses, as measured by the temperature of the cake exiting the presses, was gradually increased to about 255° F. by first slowing and then stopping the flow of cooling water to obtain operating temperatures of 180-200° F., and then by substituting steam for the water in the shaft and cages at gradually increasing pressures to achieve temperatures of 200-255° F.

Samples of the extracted oil and press residual solids were pulled at various temperature intervals as the temperatures were increased. Samples of the residual solids were pulled at two points, the first (non-rehydrated) immediately after exiting the cake-cooling screw following the final (second) pressing stage, and the second after the thus-cooled residual press solids were rehydrated to a moisture content of about 10%. The samples were assayed for ASTA colour, aerobic and anaerobic plate count, and colour stability over time using methods employed in Examples 1A and 1B.

The advantages of operating the process at a temperature above 130° F., as indicated by the temperature of the cake exiting the presses, can clearly be seen. The plate count of both the extract and the cake are progressively reduced as the temperatures are increased.

Table: *Effect of Increasing Temperatures on the Plate Count of the Extract*

Temperature Degree F	***Aerobic Plate Count***	***Anaerobic Plate Count***
80	1,900,000	790,000
130	1,700,000	800,000
150	1,700,000	660,000
170	1,600,000	590,000
175	1,500,000	425,000
180	1,300,000	380,000
190	360,000	150,000
200	300,000	200,000
215	240,000	150,000
225	190,000	65,000
235	170,000	32,000
245	69,000	8,600
255	3,800	830

Table: *Effect of Increasing Temperatures on the Plate Count of the Press Solids*

Temperature Degree F	***Aerobic Plate Count***	***Anaerobic Plate Count***
80	220,000	55,000
130	160,000	35,000
150	160,000	25,000
170	100,000	20,000
175	32,000	15,000
180	80,000	7,400
190	3,500	800
200	9,800	3,400
215	5,800	2,300
225	4,100	500
235	1,900	1,100
245	5,400	100
255	800	100

The efficiency of extraction is dramatically improved as evidenced by the progressively decreasing ASTA values and the progressively decreasing residual extractable yields of the press residual solids. It is apparent that, to achieve residual extractable yields of less than about 20% by weight of the cake, it is necessary to operate the presses at 130° F. or higher. Moreover, for obvious reasons of efficiency, temperatures above 180° F., and especially between about 180° F. and about 235° F., are greatly preferred.

Table: *Press Cake ASTA and Residual Yields at Progressively Increasing Temperatures*

Temperature Degree F	*Press Solids ASTA*	*Press Solids Residual Yield*
80	87	28.28%
130	76	16.40%
150	65	15.72%
170	61	15.72%
175	53	12.36%
180	43	13.88%
190	42	10.84%
200	44	10.72%
215	41	9.96%
225	39	9.50%
235	33	9.28%
245	32	9.00%
255	35	9.80%

Most importantly, the stability of the extract is not adversely affected and is in fact increased. The accelerated study was done according to the procedures described in Example 1B with the colours reported as a percent of the starting colour for each respective sample to adjust for the varying colour yields at the respective temperatures.

These results demonstrate that the extract produced at higher operating temperatures exhibits increased resistance to oxidative colour deterioration. This is surprising, as explained in the following.

Table: *Press Oleoresin (Extract) Stability, Accelerated, 65° C.*

Temperature Degree F	*Hour 2*	*Hour 4*	*Hour 8*	*Hour 12*	*Hour 17*
80	94%	88%	81%	73%	62%
130	94%	89%	82%	75%	64%
170	93%	89%	82%	76%	65%
225	94%	90%	82%	78%	67%
235	94%	90%	82%	77%	69%
255	95%	90%	84%	78%	72%

It is commonly believed that lipid-containing systems, when exposed to heat, will exhibit an increased rate of lipid oxidation that, once initiated, will proceed at an ever-increasing rate. (Rancidity and its Measurement in Edible Oils and Snack Foods, A Review, Robards, Kerr, and Patsalides, Analyst, February 1988, Vol 113). In fact, prior art (U.S. Pat. No. 4,681,769) claims a process for counter-current, high pressure extraction of Capsicums at less than 100° F. and less than 500 PSI for the express reason of protecting the extracted oil from oxidation.

To confirm the positive effect of high temperature treatment in more controlled conditions, a forty gram sample of hexane-extracted oleoresin paprika, with no diluents added, was heated in a beaker on a heated stir plate at 100° C. for eight and one-half hours. A control sample which was unheated, a sample pulled from the heated beaker after four hours, and a sample of the material heated for the full eight and one-half hours were dispersed on flour salt to make dispersions of 1.2% oleoresin by weight of flour salt. Two gram-portions of the dispersions were weighed into test tubes and placed in a 65° C. oven.

An initial ASTA colour was run on each dispersion and then ASTA colours were run periodically and the results were plotted versus time to determine the relative stability of the heated and unheated samples.

Table: *Stability of Non-Rehydrated Press Solids at Various Press Operating Temperatures, Expressed as Percent of Starting Colour Retained*

Temperature	***Week***	***Week***	***Week***	
Degree F	2	4	6	
80	86.7%	82.2%	85.5%	
130		89.6%	85.5%	84.6%
170		73.3%	65.3%	58.1%
225		61.7%	35.8%	32.5%
245		68.2%	31.0%	19.3%

But, very importantly, it can be seen that the press residual solids which are rehydrated immediately after exiting the second press stage of the process (Example 2) exhibit significantly increased stability relative to the non-rehydrated solids, thus overcoming the claimed disadvantages from operating at temperatures above 100° F. as set forth in U.S. Pat. No. 4,681,769.

Table: *Stability of Rehydrated Press Solids at Various Press Operating Temperatures, Expressed as Percent of Starting Colour Retained*

Temperature	***Week***	***Week***	***Week***
Degree F	2	4	6
80	90%	92%	91%
130	93%	91%	92%
170	92%	92%	91%
225	94%	93%	91%
245	95%	94%	93%

In fact, after discounting for the effect on pigment stability of increasing residual extractable yields in the press solids obtained at the lower temperatures, the carotenoid pigments in the residual solids would show enhanced stability for a given residual extractable yield. These are surprising and unexpected results and clearly overcome the supposed obstacle of operating at elevated press temperatures and pressures.

It is further surprising that the colour stability of the residual press solids is significantly improved by controlling the water activity (AW) of the solids in ranges above those suggested for the stabilization of lipid-containing systems by extensive studies and particularly by Nelson and Labuza, Water Activity and Food Polymer Science: Implications of State on Arrhenius and WLF Models in Predictina

Shelf Life, K. A. Nelson & T. P. Labuza, Journal of Food Engineering 22, 271-289 (1994). Water activity is defined as the ratio of the vapour pressure of water in a food to the vapour pressure of pure water at the same temperature.

Prior art suggests that maximum stability of lipid systems should be attained at water activities of about 0.3 with decreasing stability developing as the water activity is increased above this level. In this example we find precisely the inverse effect on stability of the carotenoid pigments for a given water activity.

In order to confirm the effect of high temperatures in the pressing operation, and to confirm the effect of added moisture, a controlled test was performed on a laboratory scale where the effect of levels of extractable yield in the cake could be controlled to eliminate the effect of variable press cake residual yields on the stability of the carotenoids. A 3,000 gram sample of ground paprika solids (175 ASTA, 9.8% extractable yield) was dried in a lab tray dryer at 100° F. for 16 hours to a moisture content of about 2%. One half of this sample was then heated in an oven at 220° F. for twenty minutes to approximate the temperature in a pressing operation according to the invention.

The other unheated sample served as a control. One hundred gram samples of each of the two materials were rehydrated at approximately 1% intervals up to about 12% moisture. The water activity AW of each was determined using a Rotronics Hygroskop DT, model DT2/1-00IV, water activity instrument. Samples were weighed into sealed test tubes, stored at ambient temperatures of about 72° F. in the dark, and the ASTA colours were determined over a period of eighteen weeks to determine the relative rates of colour degradation. The colour retained (as a percentage of the starting colour for each sample to compensate for the effect of colour dilution with the rehydration water) was plotted against time.

Table: *Percent Colour Retained of Unheated Ground Paprika at Various Water Activity Ranges*

Water Activity Aw	*Week 1*	*Week 5*	*Week 18*
0.15	74%	57%	42%
0.30	50%	45%	12%
0.40	68%	50%	43%
0.60	83%	68%	55%

Table: *Percent Colour Retained of Heated Ground Paprika at Various Water Activity Ranges*

Water Activity Aw	*Week 1*	*Week 5*	*Week 18*
0.15	66%	56%	41%
0.30	60%	50%	45%
0.40	80%	62%	57%
0.60	98%	82%	78%

It can be seen in Tables 9 & 10 that the stability of the carotenoid pigments follows almost precisely the inverse of the curve predicted by Nelson & Labuza. It can also be seen from these tables that controlled temperature (with concurrent browning) significantly enhances the stability of the carotenoids above a water activity of 0.3 and particularly in the water activity range of 0.4 to 0.6. Water activity ranges higher than 0.6 were not tested as levels marginally higher than this range will support microbial growth which is not acceptable in a dry spice product.

It can be concluded that the stability of the carotenoid pigments found in Capsicums unpredictably does not follow the commonly-accepted and predicted pattern for lipid oxidation with respect to temperature and water activity as suggested in U.S. Pat. No. 4,681,769, or in the cited literature (Nelson and Labuza, Water Activity and Food Polymer Science: Implications of State on Arrhenius and WLF Models in Predicting Shelf Life, K. A. Nelson & T. P. Labuza, Journal of Food Engineering 22, 271-289 (1994); Rancidity and its Measurement in Edible Oils and Snack Foods, A Review, Robards, Kerr, and Patsalides, Analyst, February 1988, Vol 113); describing the stability of lipid systems.

In fact, high temperature treatment, combined with rehydration of the press solids to a water activity above 0.3, preferably of 0.4 to 0.6, significantly improves stability rather than decreases it. This is a very surprising and unpredicted result.

It is well known that the lipid profile of Capsicum and its extracts, without the addition of any diluents, comprises a mixture of saturated and unsaturated fatty acids, 60-70% being unsaturated linoleic and linolenic, Lipid and Antioxidant Content of Red Pepper, Daood, Biacs, et al., Central Food Research Institute, Budapest, Hungary (1989) and The Nature of Fatty Acids and Capsanthin Esters in Paprika, Nawar et al., Journal of Food Science, Vol 36 (1971).

In fact, Daood et al suggest that "...the presence of triglycerides containing high amounts of unsaturated fatty acids may be an important factor contributing to the fading of paprika during processing and storage." The present findings are just the opposite. Without in any way being limited by theoretical considerations, it is hypothesized that the presently-discovered surprising and unpredicted inverse relationship shown (in Tables 9 & 10) between the stability of carotenoid pigments at given water activities is due to the fatty acids in the substrate being Preferentially attacked by the oxidation reaction at the low (from about 0.05 to 0.2 AW) and higher water activity ranges (above 0.3, preferably about 0.4 to 0.6 AW), thus protecting the carotenoids. At the intermediate water activity ranges (0.2 to 0.4 AW), where the lipids are best protected, the carotenoids are more readily and preferentially attacked and exhibit low resistance to oxidative degradation.

Another controlled test was conducted to demonstrate the effect of different extractable yields in the residual solid press cake. The effect of higher amounts of unsaturated fatty acids is evident from the results illustrated where fresh, refined, bleached, and deodorized soybean oil with no antioxidants was added at various percentages based on the weight of the paprika. The colour over time was compared to the untreated control in an accelerated study at 65° C. A typical Refined, Bleached, and Deodorized soy oil has a fatty acid composition of 22.3% Oleic (18:1), 51% linoleic (18:2), and 6.8% linolenic (18:3). (Riegel's Handbook of Industrial Chemistry, 9th Edition, pg 278). It can be concluded that higher levels of unsaturated fatty acids, such as oleic, linolenic, and linoleic, which are found in most vegetable oils, will improve the colour stability of the press residual solids.

Levels of extractable yield in the residual solids above about 15-20% by weight of the residual solids is undesirable as the residual Capsicum solids become difficult to handle for most uses and the efficiency of extraction is reduced, *i.e.*, less colour can be removed from the spice as the residual yield is allowed to increase by decreasing either the pressure or temperature employed.

Table: *Percent Colour Retained with Varying Amounts of Soy Oil Added to Ground Paprika*

Percent Addition	*Hour 2*	*Hour 4*	*Hour 6*	*Hour 8*
0%	65%	59%	52%	50%
5%	90%	83%	74%	72%
10%	92%	84%	75%	74%
15%	94%	87%	80%	78%
20%	96%	91%	83%	81%

It is readily apparent, comparing the results of the controlled test (Tables 9 & 10) on stability of heated vs unheated material, where oil is controlled at a constant level that, at a given added soy oil content in the press residual solids, the colour stability of the residual press solids is significantly improved when the spice, *e.g.*, a Capsicum, has been exposed to higher temperatures.

This conclusion is not readily apparent in the results shown in Table where the amount of residual vegetable oil left in the press residual solids is higher in the low temperature ranges due to the decreased efficiency of the extraction process at lower temperatures. The presence of higher amounts of residual oils there offers some protection which overshadows the increased protective effect at higher temperatures so evident in Tables.

It can therefore be concluded that much, if not all, of the protection offered by operating the presses at temperatures lower than 100° F. (as claimed in U.S. Pat. No. 4,681,769) as compared to temperatures above 100° F. is simply due to the higher residual oil levels (reduced extraction efficiency) and that, for any given residual oil content, and with rehydrated residual solids, the operating temperatures above 130° F. give superior results, not only in an increased extraction efficiency which allows for a continuous, high speed process with increased throughput rates and significantly reduced microbial activity, but most surprisingly in an increased colour stability of both the extract and the residual press solids, particularly when the press solids are rehydrated.

Comparative Example: According to Bennett U.S. Pat. No. 4,681,769,—Low Temperature and Pressure. The press solids residual yield is much higher at temperatures below 100° F. and much higher (28.3% residual yield) than disclosed in U.S. Pat. No. 4,681,769 (10-15% residual yield). In an effort to more closely model the residual yields of 10-15% (oil) in the cake as disclosed in U.S. Pat. No. 4,681,769, the feed rate for this test was set at about 95 pounds per hour, thus allowing more residence time in the press to expel more extract and to reduce the residual yield of the press residual solids to 10-15%.

Aromatherapy

Basic Principles of Aromatherapy

Essential oils are obtained by 2 main methods:

(1). Expression (also called pressing) *i.e.* cold pressed lemon oil.

(2). Distillation, either steam, water or dry.

For oils such as Camphor it is processed three times to produce the three types of oil. The first produces Brown Camphor, the second Yellow Camphor and the third White Camphor.

The other methods for extraction are:

(1). Solvent. This produces a 'Concrete', a 'Resinoid' and an 'Absolute'.

(2). Enfleurage or Pomade. This method for producing Essential oils is not used much any more, as it is an expensive and time consuming process.

A Concrete is obtained through the use of a hydrocarbon solvent to extract the Essential oil from the plant matter. This is used for the Essential oils such as Rose, Jasmine and Ylang Ylang. The Ylang Ylang concrete is approx. 80% Essential oil and 20% wax. Jasmine is approx. 50% Essential oil and 50% wax. A second extraction may also be performed to the plant matter using alcohol which produces an Absolute such as Neroli.

A Resinoid is obtained by the same method, but it is produced from resin based plants such as Amber and Frankincense. A Pomade was obtained by the use of layers of fat onto which the petals of plants such as Tuberose and Jasmine were laid out and left to dry.

The fat collected the Essential oils which were later extracted. This process has now been replaced by solvent extraction. Essential oils are found in very small quantities in many plants. For example: for every 100 kg of plant matter the following plants produce these amounts of oil:

Eucalyptus	-	3kg
Lavender	-	1.9kg
Ylang Ylang	-	1.6-2kg
Juniper	-	1/2-1.2kg
Rose	-	0.05kg

As you can see it takes a lot of plants to make a small amount of Essential oil.

Plants from different areas/countries can be more expensive (lower % of oil per plant) or cheaper (higher % of oil per plant).

Essential oils are categorized by:

(a) Their plant type *i.e.* Citrus

(b) By their Note. *i.e.* Base note

The plant types are: Citrus, floral, herb, spice, wood and resin. Although the oils can blend with any other oils, they blend better with oils of their own group, or with an oil of a similar group. *i.e.* Lemon (citrus) & Lavender (floral) and Cedarwood (wood) & Patchouli (herb). The note types are: Top, Middle and Base. Top note oils such as Neroli and Lemon evaporate and lose their aroma quickly when left open. A top note oil will last approx. 1 week if left opened. A middle note oil such as Lavender and Geranium is slightly more stable and will evaporate and lose its aroma approx. 2-3 weeks when left opened. Base note oils are the heavy oils such as Sandalwood and Patchouli which will evaporate much more slowly, taking about a month.

When you smell a blended oil you can usually pick which oils are the top, middle and base oils as the first one you notice will be a top note. The next scent you notice will be from the middle notes and the heavy, lingering scent will belong to the base note. In this way you can tell which oil is which note.

A good blend will contain at least one of each note, to add a layered effect to the blend.

Synthetic Vs. Natural Oils

Most of you will have seen cheap bottles of 'fragrant' oils or bottles of 'blended Essential oils' as well as bottles of 'Pure Essential oils'. Each one has it's own uses.

Fragrant Oils: These are the synthetic oils.

Scientists have been able to reproduce about 90% of the natural occurring Essential oil. This is what a fragrant oil is. It is the remaining 10% of the natural Essential oil that holds the therapeutic content of the oil, and that 10% is what distinguishes the two.

Fragrant oils have NO therapeutic content, they smell nice (sometimes better than the Essential oils!) but that is all that they should be used for. They work well in oil burners, in baths, in pot pourri, as a perfume and various other uses, but will have no healing effect other than to smell nice.

Blended Essential Oils: These are made from Pure Essential oils, but have been diluted with a 'Carrier oil', usually Sweet Almond, jojoba, Safflower or Apricot Kernel.

These oils are blended because Essential oils are too strong to apply directly to the skin, they must be diluted first. The Blended oil has been diluted so it is safe to apply directly onto the skin. These have therapeutic qualities, which are usually printed on the label.

They may also be blended to make them cheaper, such as Rose, Neroli, Jasmine, Chamomile and Ylang Ylang.

Pure Essential Oils: These are the Essential oils in their 100% pure state.

These oils should never be applied straight onto the skin, and many Essential oils will have safety data printed onto the label if they have any harmful effects. (there are a few exceptions to the "never put oils directly on the skin" rule... which I will go into later) When purchasing an Essential oil there are 4 ways to help tell if you are buying the pure oil:

(1) Look for the words "100%" and "Pure Essential oil". This is only a guide as many brands of fragrant oil have 100% pure written on them, and some brands of Essential oils (such as the brand I use) don't have 100% pure on the label. There are also sneaky people who use the "100% Pure Essential Oil" as their brand name....(and if the word 'fragrant' appears anywhere, chances are it isn't an Essential oil.)

(2) Smell the oils, Pure Essential oils smell like the plant it comes from. If the scent is slightly fake, very sweet or in the case of Rose really strong, it is more likely to be a fragrant oil.

(3) Look at the price, Pure Essential oils are expensive, due to the quantity of plants needed to produce the oil (for *e.g.* it takes approx. 15 roses to make 1 drop of Rose oil.) Each oil should have a different price, as some are cheaper to produce than others. As a general rule... if they are under $5 (Australian money... not sure of the prices in other countries) they are probably either blended or a fragrant oil.

(4) There are no Essential oils of Strawberry, Dewberry, Rainforest or Nanna's Garden. If something like these is packaged the same and on the same stand, then chances are all the oils are synthetic oils.

This is a guide only... but it helps to work out which ones are "fake" and which ones are "real". The store keeper may not know the difference, as they buy products from a distributor, and many haven't a clue what they are... they just know they sell well:

Aromatherapy vs. "Normal" Medicine

Many people take Aspirin for headaches although many tests have proven that Aspirin can cause stomach upsets, thin the blood, cause liver damage and contribute to anaemia. Yet people take many of these a day to relieve headaches, and think they are safe.

Antibiotics are used for infections because they kill bacteria. However there are good bacteria living in our bodies, without which our bodies can not work efficiently. Antibiotics kill off all bacteria, and many people suffer from thrush and other illnesses as a result. The other concern with antibiotics is that each time they are used, the body builds up a resistance to them, much the same way our bodies build up a resistance to the Smallpox or Measles virus when we are vaccinated. Stronger doses of antibiotics are then needed to fight the infections, and the circle continues.

Aromatherapy works on the wholistic approach. It treats the whole body at the one time. Aromatherapy has no negative side effects when used properly, and is non- addictive.

Aromatherapy Explained (Simply)

People think of Aromatherapy as being smell therapy, but this isn't the case. The essential oils can used in many different methods, I will be explaining those later... but the reason people think of them as being smell therapy, is that the blood vessels in your nose are very close the surface of your skin....so the molecules of "healing goodness" can be absorbed into the body quickly (and they do smell nice...well....some of them!) the healing effects can also pass through the skin anywhere on your body, and enter the blood stream, where they work on the troubled areas of the body. Essential oils, unlike prescription drugs, work only on those areas that are "broken" *i.e.* they do not go into healthy tissues, and start working there... they travel around your body, and look for the illnesses, and target them, leaving the rest of your body alone. It is because of this, that if you run lavender oil into your fingertip, it will help your infected toe...even though the oil never touches the actual toe. (and your headache, and cut finger!!!).

Using Essential Oils

There are 8 main ways to use the Essential oils, they are as follows:

Inhalation:

(1). Straight from the bottle—Headache, memory booster, nausea etc.

(2). Oil burner—Kill airborne bacteria (prevents colds spreading to others), insomnia, stress etc. Put 3-5 drops in a water filled well of the oil burner, and replace as needed. NEVER burn Essential oils without water... it damages them

(3). Drops on a tissue (carry with you or place under pillow etc.)—colds, coughs, migraine etc. 1-2 drops on a tissue will be ample

(4). Drops in sink/bowl of hot water—colds, respiratory infection, catarrh (runny nose) etc. stick your head over the sink/bowl, with eyes closed and inhale the vapour. No more than 5 drops to half a sink full of water (which is all you need)

Bat:

(1). Hot/warm bath—colds, muscle cramp, stiffness etc. 10 drops Maximum

(2). Cold bath—Fever. 10 drops Maximum

(3). Foot bath—Athlete's foot, blisters, aching feet etc. Use a bucket or bowl big enough to comfortably put your feet in. 5 drops Maximum

(4). Shallow bath—Thrush, Piles etc. a bath deep enough to cover the problem area. 5 drops maximum

If using peppermint oil, you may want to use only half the recommend drops.... as my Aromatherapy teacher said "It runs amok amongst your genitals" it has a cooling effect that you may not wish to have touching sensitive areas.

You will need to swish the water around. There is a product called Solubalizer that may be useful. It makes the oil dissolve in water, although the same effect can be to mix the oil with isopropyl alcohol (rubbing alcohol) or vodka, which will make it water soluble. 5 drops to 1tsp alcohol. I have also heard that the same effect can be achieved with milk, but I am not sure.

Massage:

(1). Massage diluted oil onto effected area - varicose veins, strains, constipation Muscle aches etc.

(2). Massage diluted oil all over body - Stress, insomnia, anxiety etc.

Both, use a 3% dilution for normal skin, 1% for face or sensitive skin (explained further below).

Internally: Most Essential oils are toxic and should NEVER be taken internally. It is safest to assume that ALL Essential oils are toxic and therefore none should be taken internally.

Many Aromatherapy books suggest a mouth wash or a gargle for gum problems or throat infections, however it is dangerous and other methods work just as well.

There are many different species of the same oil, for example Birch has 2 varieties, white and sweet. White birch is non-toxic, but sweet birch is fatally toxic. I would not use ANY Essential oils internally, especially in their pure state, as there are many other methods of application which are much safer. {as I was saying in a previous section, even rubbing an oil on your toe will help a sore throat, so it isn't worth ingesting something that is potentially harmful, although I have gargled with 2 drops Bergamot oil mixed with a cup of warm water}

Directly To The Skin: There are only four Essential oils which may safely be applied directly onto the skin an all should have a patch test done first. All essential oils are acidic, have you ever seen what they do to plastic?

Place a drop of the oil onto the back of your wrist, cover with a Band-Aid and leave for 1 hour. If no irritation has occurred you may use it. If irritation does occur, bathe the area in cool water and dilute the oil with a carrier oil. Only the four Essential oils listed below may be applied to skin directly, and only onto an effected area *i.e.* a cut or burn. Do not use as a massage oil, or slather it all over your skin:

Lavender, Sandalwood, Tea tree and some say Lemon, ylang ylang or Chamomile.

There is no need to apply any essential oil directly onto the skin, an essential oil blended 3% into a carrier oil will have exactly the same healing properties as a full strength oil... so not only do you run the risk of burning the skin by applying them neat, you are also wasting oil:)—Scrooge McObsidian here:)

(1). Drops of oil in cold water for cold compress - fever, swelling etc.

(2). Drops of oil in hot water for hot compress - Headache, period pain etc.

To make a compress, half fill a bowl or sink of either hot or cold water, and add 3-5 drops of essential oil.... soak the cloth in this water for a few moments, wring out, and apply to the effected area.

Perfume:

(1). Use same dilution as massage oil, use a carrier oil of Apricot kernel or other light oil. Dab behind ears, wrists etc.

Skin/Hair Tonic:

(1). Use this when an oil is not suitable, for example on an oily scalp, or to dry out a cut. use 5 drops essential oil into a teaspoon of Isopropyl alcohol (rubbing alcohol) or vodka

Dilutions:

Babies 0-12 months—1 drop of Rose or Lavender or chamomile in 1 teaspoon of carrier oil, or in a bath.

Infants 1-5 years - 2-3 drops of Rose, Lavender, Chamomile, Sandalwood, Tangerine, Ylang Ylang or Neroli in 1 teaspoon of carrier oil or in a bath.

Children 6-12 years -Use as for adults, but half the concentration.

3% dilution:

For 100 mls of carrier oil use 60 drops of Essential oil.

For 25 mls of carrier oil use 15 drops of Essential oil.

For 5 mls of carrier oil (1 teaspoon) use 3 drops of Essential oil.

1% dilution:

For 100 mls of carrier oil use 20 drops of Essential oil.

For 25 mls of carrier oil use 5 drops of Essential oil.

For 5 mls of carrier oil (1 tsp) use 1 drop of Essential oil.

For normal use do not exceed 3% Essential oil dilution.

For use on face or other sensitive skin use a 1% Essential oil dilution.

Pregnancy - use half the dilution and none of the contraindicated oils.

Contraindications

Aromatherapy oils are concentrated and should not be applied to the skin or taken internally. There are also other times when Essential oils should be treated with caution or not be used at all, these are called Contraindications. The following is a list of the contraindications for the various Essential oils.

Oils not to be Used at All: These oils are dangerous for anyone except a qualified Aromatherapist to use. They are either extremely toxic or cause severe skin irritation even in a diluted state. These oils are: Bitter Almond, Arnica, Boldo, Broom, Buchu, Calamus, Camphor (brown & Yellow), Cassia, Chervil, Cinnamon (bark), Costus, Deertongue, Elecampane, Fennel (bitter), Horseradish, Jaborandi, Melilotus, Mugwort, Mustard, Oregano, Pennyroyal, Pine (dwarf), Rue, Sage (common), Santolina, Sassafras, Savine, Savoury, Tansy, Thuja, Thyme (red), Tonka, Wintergreen, Wormseed and Wormwood.

Oils that should be Used in Small Doses for no Longer Than 2 Weeks: These oils are fairly toxic or may have side effects such as nausea, vomiting and headaches and should be used with extreme caution:

Ajowan, Anise star, Aniseed, Basil (exotic), Bay laurel, Bay (west indian) Calamintha, Camphor (white), Cascarrilla bark, Cassie, Cedarwood (virginian), Cinnamon (leaf & bark), Clove (bud), Coriander, Eucalyptus, Fennel (sweet), Hops, Hyssop, Juniper, Nutmeg, Parsley, Pepper (black), Pine, Sage (spanish), Tagetes, Tarragon, Thyme (white), Tuberose, Tumeric, Turpentine and Valerian.

Oils that Irritate the Skin if Used in High Concentration: These oils should be used in half the recommended dilution, and no more than 3 drops in a bath. These oils are:

Ajowan, Allspice, Aniseed, Basil (sweet) Black pepper, Borneol, Cajeput, Caraway, Cedarwood (Virginian), Cinnamon (leaf), Clove (bud), Cornmint, Eucalyptus, Garlic, Ginger, Lemon, Parsley, peppermint, Pine (needle, Scotch & Longleaf), Thyme (white) and Tumeric.

Oils that Cause Irritation on Sensitive Skin: These oils may cause eczema or dermatitis with people who have very sensitive skin. (For people with senstive skin always test the oil on the back of your wrist, and leave for an hour. If irritation occurs bathe area with cold water and try a weaker concentration.) These oils are:

Aniseed, Basil (french), Bay laurel, Benzoin, Bergamot, Cade, Cajeput, Cananga, cedarwood (virginian), Chamomile (Roman and German), Citronella, Garlic, Geranium, Ginger, Hops, Jasmine, Lemon, Lemongrass, Lemon balm (Melissa), Litsea cubeba, Loveage, Mastic, Mint (pepper & spear), Orange, Peru balsam, Pine (scotch & longleaf), Styrax, Tea tree, Thyme (white), Tolu balsam, Tumeric, Turpentine, Valerian, Vanilla, Verbena, Violet, Yarrow and Ylang Ylang.

Oils that are Phototoxic: These are oils which can cause the skin to darken if exposed to direct sunlight. Do not use these oils at all if the area will be exposed to sunlight. These oils are:

Angelica root, Bergamot, Cumin, Ginger, Lemon, Lime, Loveage, Mandarin, Orange and Verbena.

Oils that should be Avoided during Pregnancy: Due to the effects of these oils on the reproductive organs, and the sensitivity of the foetus, certain oils should not be used at all during pregnancy. These oils are:

Ajowan, Anjelica, Anise star, Aniseed, Basil, Bay laurel, Calamintha, Cedarwood (all types), Celery seed, Cinnamon (leaf),

Citronella, Clary sage, Clove, Cumin, Cypress, Fennel (sweet), Hyssop, Jasmine, Juniper, Labdanum, lovage, Marjoram, Myrrh, Nutmeg, Parsley, Penyroyal, Peppermint, Rose, Rosemary, Snakeroot, sage, Tarragon and Thyme (white)

Oils that should be Avoided with High Blood Pressure: These oils should not be used:

Black pepper, Hyssop, Lemon, Lemongrass, Nutmeg, Rosemary, Sage (spanish & Common) and Thyme.

Oils that should be avoided with low blood pressure. - These oils should not be used:

Chamomile (Roman & German), Lemon balm, Lavender (true), Marjoram (sweet) and Ylang ylang.

The oil that should be avoided for diabetes. Anjelica.

The oil that should be avoided for kidney problems. Juniper.

Oils that should be not be Used with Homeopathic Treatments: These oils are not to be used by anyone receiving homeopathic treatment:

Black pepper, camphor, eucalyptus and peppermint.

Oils that should be Avoided with Alcohol Consumption: These oils will increase the effects of alcohol:

Aniseed, clary sage and fennel.

The oil that should be avoided with depression. Basil.

Oils that should be Avoided with Epilepsy: These oils should not be used:

Fennel, Hyssop, Sage and Rosemary.

The Common Essential Oils

The following information has been gathered from many books over many years. Some oils work by promoting balance, which is why they are listed as working well with things such as oily skin and dry skin. What this means is that if your skin is oily, it will help dry it out, if it is dry it will help add moisture. In other words, it will balance.

Different sources will give different properties to the Essential oils - some may even contradict others. It is for this reason that it is best to leave Aromatherapy to trained professionals.

The following information is not intended to prescribe or diagnose in any way. It is not meant to be a substitute for professional medical advice/assistance. The intention of this list is to give information on the historical usage of Aromatherapy. Please consult your doctor or other health professional if you have any medical complaints. Do not attempt to self diagnose or self prescribe herbal treatments for yourself or others. The Author (Obsidian) shall not be responsible for any misuse or abuse of this information.

Aniseed

Botanical Name: Pimpinella Anisum

Note - Top to middle

Type - Spice

Family - Umbelliferae

Part - Herbs/seeds

Extraction - Steam Distillation

Aroma: Sweet and spicy, pungent, Licorice like, very warming.

Blends well With: Amyris, Bay, Cardamon, Caraway, Cedarwood, Coriander, Dill, Fennel, Galbanum, Mandarin, Petitgrain, Rose wood.

Contraindications: Sensitive skin, Pregnancy, Over use can cause sluggishness, drowsiness and dizziness. In extreme cases can cause cerebral congestion and circulatory problems.

Properties: Carminative, Antiseptic, Antispasmodic, Decongestant, Parasiticide, Antiemetic, Aphrodisiac, Cardiac, Digestive aid, Diuretic, Expectorant, Galactagogue, Insecticide, Laxative, Parturient, Pectoral stimulant, Stomachic.

Uses: Respiratory problems, Digestive problems, Flatulence, Indigestion, Period pains, Stimulate lactation, Lice, Scabies, Muscular aches and pains, Rheumatism, Bronchitis, Coughs, Colic, Cramp, Colds, Stimulate the mind, Dyspepsia, Nausea, Vomiting, Stimulating peristalsis, Oliguria (low quantity of urine), Stimulates cardiac fatigue, Lice, Scabies, Infectious skin diseases.

Cautions: Not generally recommended for use in the home as prolonged or over use may be harmful.

Basil (Sweet)

Botanical Name: Omicum Basilicum

Note - top

Type - Herb

Family - Labiatae

Part - Flowering top

Extraction - Steam Distillation

Aroma: Very clear, sweet and slightly spicy.

Blends well With: Bergamot, Black pepper, Citronella, Chamomile, Clary Sage, Geranium, Hyssop, Lavender, Lemongrass, Lime, Marjoram, Melissa, Neroli, Peppermint, Rose, Sandalwood, Verbena.

Contraindications: Pregnancy, Sensitive skin, Do not use to excess.

Properties: Soothing, Uplifting, Antiseptic, Insect repellent, Digestive aid, Antispasmodic, Emmenagogue, Analgesic, Antidepressant, Antispasmodic, Antivenomous, Aphrodisiac, Bactericide, Carminative, Cephallic, Expectorant, Febrifuge, Galactagogue, Insecticide, Nervine, Stomachic, Sudorific, Tonic, Restorative, Stimulant, Vermifuge, Effects adrenal cortex, Imitates oestrogen, Minimise uric acid in muscles, stimulates blood flow, Helps expel afterbirth, Cleanse intestines and kidneys.

Uses: Fatigue, Depression, Respiratory infections, Colds, Cough, Bronchitis, Asthma, Sinusitis, Fever, Gout, Indigestion, Insect bites and stings, Breast engorgement, Fainting, Flatulence, Insomnia, Muscular aches and pains, Rheumatism, Anxiety, Migraine, Nervous tension, Sharpening senses, Nerve tonic, Concentration, Hysteria, Headaches, Temporary paralysis, Nasal polyps, Earache, Allergies, Bronchitis, Emphysema, Flu, Whooping cough, Catarrh, Vomiting, Gastric spasm, Nausea, Dyspepsia, Hiccups, Menstrual problems, Scanty periods, Fertility problems, Fatigue, Stress, Poor circulation.

Bay

Botanical Name: Laurus Nobilis

Note - top

Type - Spice

Family - Lauraceae

Part - Leaves and twigs

Extraction - Steam Distillation

Aroma: Sweet and spicy, similar to Cinnamon

Blends well With: Citrus Oils, Cedarwood, Clary Sage, Coriander, Cypress, Eucalyptus, Ginger, Hyssop, Juniper, Lavender, Lemon, Marjoram, Myrtle, Orange, Rose, Rosemary, Thyme, Ylang Ylang.

Contraindications: Sensitive skin, Pregnancy, Irritate mucous membranes.

Properties: Analgesic, Antineuralgic, Antiseptic, antispasmodic, Aperitif, Astringent, Cholagogue, Diuretic, Emmenagogue, Febrifuge,

Hepatic, Insecticides, Parturient, Stimulant, Stomachic, Sudorific, Tonic, Mildly narcotic, Warms emotions, Reproductive tonic, Uplifting.

Uses: Colds, Flu, Bronchitis, Digestive aid, Dyspepsia, Flatulence, Appetite stimulant, Settles stomach, Liver and Kidney tonic, Rheumatism, Aches and pains, Sprains, Fever, Infectious disease, Scanty periods, Speeds up childbirth, Ear infections, Dizziness, Restores balance, Minor respiratory problems, Indigestion.

Cautions: People prone to sensitive or allergy prone skin should not use Bay oil. Use in moderation and never undiluted.

Bergamot

Botanical Name: Citrus Bergamia

Note - top

Type - Citrus

Family - Rutacaea

Part - Peel

Extraction - Expression

Aroma: Light, delicate, refreshing, something like Orange and lemon with slight floral undertones.

Blends well With: Basil, Cardamon, Chamomile, Coriander, Cypress, Eucalyptus, Geranium, Juniper, Jasmine, Lavender, Lemon, Marjoram, Mimosa, Myrtle, Neroli, Palmarosa, Patchouli, Petitgrain, Sandalwood, Ylang ylang.

Contraindications: Phototoxic, sensitive skin.

Properties: Antiseptic, Parasiticide, Antidepressant, Analgesic, Antispasmodic, Carminative, Cicatrisant, Cordial, Deodorant, Digestive, Expectorant, Febrifuge, Insecticide, Sedative, Stomachic, Tonic, Vermifuge, Vulnerary, Keeps pets away from plants, Uplifting, Vitalising, Anti-inflammatory,

Uses: Dandruff, Urinary and Respiratory infections, Skin infections, Throat and mouth infections, Scalp and skin, Psoriasis, Acne, Ulcers, Stress related conditions, Depression, Coldsores, Shingles, Insomnia, Anxiety, Stress, Bad breath, Deodorant, Worms, Intestinal parasites, Colic, Appetite stimulant, Vaginal infections, Flatulence, Boils, Eczema, Varicose Ulcers, Wounds, Tonsillitis, Colds, Fever, Flu, Abscesses, Cystitis, Digestive aid, Eczema, Insect repellent, Oily skin, Infectious diseases, PMS, Dyspepsia, Flatulence, Colic, Indigestion, Gallstones, Anorexia, Breathing difficulties, TB, Chickenpox, Uterine Tonic, STDs, Gastro-intestinal spasm, Cellulite, Sore throat, Bronchitis.

Black Pepper

Note - Middle

Type - Spice

Family - Piperaceae

Part - fruit

Extraction -Distillation

Aroma: Sharp and Spicy

Blends well With: Basil, Bergamot, Cedarwood, Cypress, Frankincense, Geranium, Grapefruit, Juniper, Lemon, Marjoram, Palmarosa, Rosemary, Sandalwood, Ylang Ylang.

Contraindications: Kidney Problems, Sensitive Skin.

Properties: Analgesic, Antiemetic, Antiseptic, Antispasmodic, Aphrodisiac, Cardiac, Carminative, Detoxicant, Digestive, Diuretic, Febrifuge, Laxative, Rubefacient, Stimulant, Stomachic, Tonic, Helps frustration, Warms the heart with there is indifference, Dilates local blood vessels, Stimulates peristalsis, Stimulates circulation, Helps form new blood cells, Antidote for fish and mushroom poisoning, Reduces fat, Increases flow of Saliva, Stimulates appetite, Expels wind, Fortifying stomach, Expels toxins.

Uses: Muscle aches and pains, Tired and aching limbs, Muscle stiffness, Rheumatoid Arthritis, Temporary Paralysis, Vomiting, Bowel problems, Anaemia, Respiratory illnesses, Fever (Small amounts).

Cajuput

Note - top

Type - Herb

Family - Myrtacaea

Part - leaves and twigs

Extraction - Distillation

Aroma: Sweet, herbaceous and rather penetrating

Blends well With: Angelica, Bergamot, Birch, Cardamon, Cedarwood, Clove, Eucalyptus, Geranium, Immortelle, Lavender, Myrtle, Niaouli, Nutmeg, Pine, Rose, Rose wood, Thyme.

Contraindications: Use with caution, sensitive skin, irritate mucous membranes.

Properties: Analgesic, Antidontalgic, Antineuralgic, Antirheumatic, Antiseptic, Antispasmodic, Balsamic, Cicatrisant, Decongestant,

Expectorant, Febrifuge, Insecticide, Pectoral, Stimulant, Sudorific, Vermifuge, Clears the mind, Balances mind and body, Imitates Oestrogen.

Uses: Respiratory tract infections, Fever, Colds, Pharyngitis, Laryngitis, Bronchitis, Asthma, Ease Chronic Pulmonary Disease, Colic, Inflammation of the intestines, Enteritis, Dysentery, Gastric spasm, Nervous vomiting, Intestinal Parasites, Urinary infections, Cystitis, Urethritis, Neuralgia, Headaches, Toothache, Earache, Gout, Chronic rheumatism, Muscle Stiffness, Muscle aches and pains, Menopausal problems.

Cedarwood

Note - Base

Type - Wood

Family - Cupressacae/Pinaceae

Part - Wood

Extraction - Distillation

Aroma: Woody, reminiscent of sandalwood but slightly "drier" and almost pine tones.

Blends well With: Benzoin, Bergamot, Black pepper, Cajuput, Cinnamon, Cypress, Frankincense, Ginger, Jasmine, Juniper, Lavender, Lemon, Linden, Myrrh, Neroli, Patchouli, Pine, Rose, Rosemary, Sandalwood, Vetiver, Ylang Ylang.

Contraindications: Sensitive Skin, Pregnancy.

Properties: Calming, Soothing, Antiseptic, Astringent, Diuretic, Emollient, Expectorant, Fungicide, Insecticide, Sedative, Tonic, Better for chronic (long standing) problems than acute (recent) ones, Balances body and mind, Insect repellent, Aphrodisiac.

Uses: Nervous Tension, Anxiety, Meditation, Bronchitis, Coughs, Catarrh, Excess phlegm, Cystitis, Genito-Urinary tract problems, Kidney problems, Rheumatism, Arthritis, Oily skin, Acne, Clearing scabs, Dermatitis, Psoriasis, Hair tonic, Seborrhoea, Dandruff, Alopecia, Softening skin (especially when mixed with Cypress and Frankincense), Respiratory infections, Glandular problems, Stress, Insect bites and stings, Dermatitis, Eczema, Itching, Coughs, Fungal infections, Hair loss, Ulcers, Bronchitis, Catarrh, PMT, Tiredness, Vaginal infections, Urinary tract infections, Congestion, Sinusitis.

Chamomile

Note -Middle

Type - Flower

Family - Compositae

Part - Flowers

Extraction - Distillation

Aroma: Fruity, apple-like.

Blends well With: Angelica, Basil, Benzoin, Bergamot, Clary sage, Geranium, Jasmine, Lavender, Lemon, Marjoram, Neroli, Palmarosa, Patchouli, Rose, Star Anise, Ylang Ylang.

Contraindications: Pregnancy (early Months).

Properties: Emmenagogue, Analgesic, Antiallergenic, Anticonvulsive, Antidepressant, Antibacterial, Antiemetic, Antiphlogistic, Antipruritic, Antirheumatic, Antiseptic, antispasmodic, Carminative, Cholagogue, Cicatrisant, Digestive, Diuretic, Emollient, Febrifuge, Hepatic, Nervine, Sedative, Splenetic, Stomachic, Sudorific, Tonic, Vermifuge, Vulnerary, Soothing, Calms the mind, Regulates menstruation, Stimulates growth of white corpuscles.

Uses: Anxiety, Tension, Anger, Fear, Promotes relaxation, Gives patience and peace, Allay worry, Insomnia, Muscular pain, Lower back pain, Headache, Neuralgia, Toothache, Earache, Menstrual problems, Period pain, PMT, Menopause, Soothes stomach, Gastritis, Diarrhoea, Colitis, Peptic Ulcers, Vomiting, Flatulence, Irritated bowels, Liver problems, Jaundice, Genito-urinary problems, Anaemia, Burns, Blisters, Inflamed wounds, Ulcers, Boils, Dermatitis, Acne, Herpes, Psoriasis, Broken Capillaries, Improves skin elasticity, Itching, Puffiness, Strengthen tissues, Skin cleanser, Hair tonic, Cramps, Nappy Rash, Nervous Tension, Neuralgia, Bleaching hair, Cracked nipples, Children's tantrums, Fevers, Insomnia, Mastitis.

Cinnamon

Note - Base

Type - Spice

Family - Lauracaea

Part - Bud/Bark/leaf

Extraction - Distillation

Aroma: Spicy, Sharp, Sweet and Musky.

Blends well With:Benzoin, Cardamon, Clove, Coriander, Eucalyptus, Frankincense, Galbanum, Ginger, Grapefruit, Lavender, Lemon, Mandarin, Orange, Pie, Rosemary, Thyme.

Contraindications: (Leaf is safer as Bud and Bark can irritate skin more) Use with care in small doses, Pregnancy, High doses could cause convulsions.

Properties: Anaesthetic, Antidontalgic, Antiseptic, Antiputrefative, Antispasmodic, Aphridesiac, Astringent, Cardiac, Carminative, Emmenagogue, Escharotic, Haemostatic, Insecticide, parasiticide, Sialogue, Stimulant, Stomachic, Vermifuge, Stimulates tears, Stimulates saliva and mucous, Stimulates secretion of gastric juices, Stimulates Circulatory system.

Uses: Respiratory infections, Colds, Flu, Breathing difficulty, Fainting, Infectious diseases, Viral infections, Intestinal infections, Digestive spasm, Asthenia, Dyspepsia, Colitis, Flatulence, Gastralgia, Diarrhoea, Nausea, Vomiting, Cholera, Typhoid, Period pain, Scanty menstruation, Leucorrhoea, Impotence, Muscle spasm, Rheumatism, Insect bites, Respiratory and digestive problems, Lice, Scabies, Tooth and Gum care, Warts, Wasp stings, Poor circulation, Rheumatism, Anorexia, Colitis, Aids childbirth, Frigidity, Cough, Depression, Flu.

Citronella

Note - Top

Type - Citrus

Family - Graminae

Part - Grass

Extraction - Distillation

Aroma: Sweet and Lemony

Blends well With: Bergamot, Cajuput, Cedarwood, Eucalyptus, Geranium, Lavender, Lemon, Mandarin, Neroli, Orange, Peppermint, Petitgrain, Pine, Sage, Ylang Ylang.

Contraindications: May irritate Sensitive Skin, Pregnancy.

Properties: Antidepressant, Antiseptic, Deodorant, Insecticide, Parasiticide, Tonic for heart and nervous system, Stimulant, Clearing mind, uplifting.

Uses: Insect repellent, Excessive perspiration, Oily skin, Cold, flu, Minor infections, Fatigue, Headache, Migraine, Neuralgia, Tired feet, Rheumatic aches and pains, Softens skin.

Clary Sage

Note - Top to Middle

Type - Herb

Family - Labiatae

Part - Herb/Flowering tops and leaves

Extraction - Distillation

Aroma: Heavy herby and nutty.

Blends well With: Angelica, Basil, Bay, Bergamot, Cardamon, Cedarwood, Citronella, Coriander, Cypress, Frankincense, Geranium, Grapefruit, Jasmine, Juniper, Lavender, Lemon, Lime, Myrtle, Petitgrain, Rose, Sandalwood, Ylang Ylang.

Contraindications: Not before driving or operating heavy machinery, Not with Alcohol consumption, Large doses can cause headaches, Pregnancy, Epilepsy

Properties: Anticonvulsive, Antidepressant, Antiphlogistic, Antiseptic, Antispasmodic, Antisudorific, Aphrodisiac, Balsamic, Carminative, Deodorant, Digestive, Emmenagogue, Hypotensive, Nervine, Parturient, Sedative, Stomachic, Tonic, Uterine, hormone balancing, Kidney tonic, Soothing, Uplifting, Balance, Encourages hair growth, Astringent, Anti-inflammatory, Antibacterial, Powerful muscle relaxant.

Uses: Nervous tension, Anxiety, Uterine problems, Scanty periods, PMT, Menstrual cramps, Stress, Fertility, Digestive problems, Flatulence, Gastric spams, Headaches, Migraine, Tension, cramp, Excessive perspiration, TB, Asthma, Sore throats, Throat infections, Brings vigour after illness, Convalescence, Overcoming drug addictions, Depression and hopelessness, Cell regenerate, Scalp problems, Dandruff, Oily hair, Inflamed and puffy skin, Hypertension, Colds, Menstrual problems, Dry skin, Digestive problems.

Clove

Note -Base

Type - Spice

Family - Myrtaceae

Part - (Tree) Bud or Leaf

Extraction - Distillation

Aroma: Spicy, strong.

Blends well With: Basil, Benzoin, Black pepper, Cinnamon, Citronella, Grapefruit, Lemon, Nutmeg, Orange, Peppermint, Rosemary.

Contraindications: (Highly irritant), Sensitive skin, Pregnancy, Not for use with children

Properties: Analgesic, Anaesthetic, Antidontalgic, Antiemetic, Antineuralgic, Antiseptic, Antispasmodic, Aperitif, Aphrodisiac, Carminative, Caustic, Cicatrisant, Disinfectant, Insecticide, Parturient, Splenetic, Stimulant, Stomachic, Uterine, Vermifuge, Expectorant, Antihistamine, Aids digestion, Antidepressant, Parasiticide, Tonic for the Kidneys; spleen and stomach, Antibacterial.

Uses: Dyspepsia, Gum infections, Toothache, Bronchitis, Catarrh, Scabies, Athletes foot, Cold, Flu, Tinea, Acne, Bruises, Burns, Cuts, Ulcers, Wounds, Arthritis, Rheumatism, Sprains, Asthma, Colic, Nausea, Minor infection, Diarrhoea, Flatulence, General weakness, Muscular aches and pains, Tension, Scar tissue, Memory, Lethargy, Vomiting, Intestinal spasm, Halitosis, Headache, Respiratory problems, TB, Impotence, Frigidity, Helps child birth pains, Lupus.

Cypress

Note - Middle to base

Type - Wood

Family - Cupressaceae

Part - (Tree) Leaves and cones

Extraction - Distillation

Aroma: Woody and spicy, but refreshing and clear

Blends well With: Benzoin, Bergamot, Clary Sage, Juniper, Lavender, Lemon, Linden, Orange, Pine, Rosemary, Sandalwood.

Contraindications: High Blood pressure, Pregnancy,

Properties: Antiseptic, Astringent, Aids healing, Insecticide, Vaso-constricting, Tonic for circulatory system and liver, Cicatrisant.

Uses: Urinary problems, Fluid retention, Excessive perspiration, Diarrhoea, Menorrhagia, Tired aching legs, Oily skin, Rheumatism, Back ache, Haemorrhage, Swelling, Nasal congestion, Cold, Flu, Scars, Wounds, Haemorrhoids, Varicose veins, Asthma, Cellulitis, Menstrual cramp, Poor circulation, Spasmodic coughs, Dysmenorrhoea, Stress, Nervous tension, Acne, Eczema, Gum disorders, Incontinence, Menopausal problems, Nose bleeds, Whooping cough.

Eucalyptus

Note - Top

Type - Wood

Family - Eucalyptus globulus

Part - (Tree) leaves

Extraction - Distillation

Aroma: Clear, sharp and piercing

Blends well With: Benzoin, Cajeput, Cedarwood, Coriander, Cypress, Juniper, Lavender, Lemon, Lemongrass, Marjoram, Melissa, Peppermint, Pine, Rosemary, Star Anise, Teatree, Thyme.

Contraindications: Sensitive skin, Kidney problems, Toxic if taken internally, Heart problems, Not for children, High Blood pressure, Pregnancy, Epilepsy, Do not use if you are on any Homoeopathic medication.

Properties: Analgesic, Antirheumatic, Antipholgistic, Antiseptic, Antispasmodic, Antiviral, Antibacterial, Balsamic, Cicatrisant, Decongestant, Deodorant, Depurative, Diuretic, Expectorant, Febrifuge, Hypoglycemiant, Insecticide, Rubefacient, Stimulant, Vermifuge, Vulnerary, Parasiticide, Cools and soothes, Strengthens nervous system,

Uses: Respiratory complaints, Croup, Bronchitis, Ringworm, Insect bites, Shingles, Chicken pox, Herpes, Muscle aches and pains, Dysentery, Hay fever, Burns, Throat infections, Colds, Chest infections, Acne, Asthma, Boils, Cold sores, Cuts, Fever, Flu, Lice, Laryngitis, Rheumatism, Skin infections, Sore throat, Urinary infections, Aids concentration, Catarrh, TB, Sinusitis, Migraine, Typhoid, Diphtheria, Malaria, Cystitis, Diarrhoea, Gall stones, Nephritis, Gonorrhoea, Diabetes, Haemorrhage, Neuralgia, Pyorrhoea.

Fennel

Note - Top to middle

Type - Herb

Family - Umbelliferae

Part - (Herb) seed

Extraction - Distillation

Aroma: Floral, herbal and spicy

Blends well With: Basil, Geranium, Lavender, Lemon, Rose, Rosemary, Sandalwood.

Contraindications : Pregnancy, Epilepsy, Toxic if used in large doses, not for babies.

Properties : Antiphlogistic, Antiseptic, Antispasmodic, Aperitif, Carminative, Detoxicant, Diuretic, Emmenagogue, Expectorant,

Galactagogue, Insecticide, Laxative, Resolvent, Splenetic, Stimulant, Stomachic, Sudorific, Tonic, Vermifuge, Reduces lactobacillus, Has an action similar to Oestrogen, Stimulates lactation, Reduces the toxic effect of alcohol, Body cleansing and detoxifying.

Uses : Kidney Stones, Menstrual problems, PMT, Respiratory problems, Urinary tract infections, Flatulence, colic, Menopausal problems, Diuretic, Gout, Liver problems, Children's complaints, Intestinal parasites, Nausea, Vomiting, Indigestion, Cellulitis, Obesity, Odema, Rheumatism, Asthma, Bronchitis, Anorexia, Bruises, Dull skin, Oily skin, Alcohol poisoning, Appetite reduction, Arthritis, Colitis, Constipation, Fluid retention, Mouth and gum problems, Hangovers, Stomach ailments, Eases digestion, Hiccups, Whooping cough, Scanty periods.

Frankincense

Note - Middle to base

Type - Resin

Family - Burseraceae

Part - (Tree) Bark

Extraction - Distillation

Aroma: Woody, spicy, and a hint of lemon

Blends well With: Basil, Black Pepper, Cedarwood, Cinnamon, Citrus oils, Galbanum, Geranium, Ginger, Grapefruit, Lavender, Orange, Melissa, Myrrh, Neroli, Patchouli, Pine, Sandalwood, Vetiver.

Contraindications: Pregnancy.

Properties: Uterine tonic, Astringent, Blood coagulant, Calming, Deepens breathing, Induces sweating, Pulmonary antiseptic, Calming, Clears the lungs, Anti-inflammatory.

Uses: Bronchitis, Laryngitis, Asthma, anxiety, Stress, Emotional upsets, Urinary tract infections, Cystitis, Uterine tonic, Wounds, Mature skin, Blemishes, Dry skin, Scars, Catarrh, Dysmenorrhoea, Cold, Flu, Nervous tension, Acne, Coughs, Laryngitis, Meditation, Menstrual problems, PMS, Menorrhagia, Respiratory conditions, Scar tissue, Improves skin tone, genito-urinary infections, Catarrh, Nephritis, Genital infections, Uterine haemorrhage, Post natal depression, Calms during labour, Breast inflammation, Dyspepsia, Belching, Wrinkles, Oily skin, Ulcers, Carbuncles.

Geranium

Note - Middle

Type - Floral

Family - Geraniacea

Part - (plant) Flowers and leaves

Extraction - Distillation

Aroma: Sweet and heavy, a little like rose with a hint of mint.

Blends well With: Angelica, Basil, Bay, Bergamot, Carrot Seed, Cedarwood, Citronella, Clary Sage, Grapefruit, Hyssop, Jasmine, Lavender, Lemon, Lime, Mandarin, Marjoram, Neroli, Orange, Patchouli, Petitgrain, Rose, Rosemary, Sandalwood, Tea Tree.

Contraindications: Sensitive skin, Pregnancy.

Properties: Analgesic, Anticoagulant, Antidepressant, Aphrodisiac, Antiseptic, Anti-fungal, Anti-inflammatory, Astringent, Cicatrisant, Cytophylactic, Diuretic, Deodorant, Haemostatic, Hypoglycemiant, Insecticide, Styptic, Tonic, Vasoconstrictor, Vulnerary, Cleanses the body of toxins, Stimulates lymphatic system, Clears digestive mucous, Balances Sebum, Skin cleanser, Improves flow of blood.

Uses: Anxiety, Depression, Balances mind, Stress, PMT, Menorrhagia, Lack of vaginal secretions, Inflammation, Breast Congestion, Congested system, Jaundice, Kidney stones, Gallstones, Diabetes, Urinary infections, Fluid retention, Swollen ankles, Throat infections, Mouth Infections, Neuralgia, Gastritis, Colitis, Insect Repellent, Eczema, Shingles, Herpes, Ringworm, Chilblains, Oily skin, Sluggish skin, Congested skin, Cuts, Menopausal conditions, Cellulitis, Mastitis, Emotional problems, tonsillitis, Burns, Muscular aches and pains, Respiratory conditions, Rheumatism, Swelling, Ulcers, Apathy, Acne, Bruises, Broken capillaries, Dermatitis, Haemorrhoids, Lice, Mature skin, Mosquito repellent, Wounds, Engorgement of breasts, Poor circulation, Sore Throat, Nervous tension, Abscesses, Boils, Bronchitis, Indigestion, Insect bites.

Jasmine

Note - Middle to Base

Type - Floral

Family - Jasminacaea

Part - Flowers

Extraction - Enfleurage/Solvent Extraction

Aroma: Sweet and flowery.

Blends well With: Bergamot, Citrus oils, Chamomile, Clary Sage, Coriander, Frankincense, Geranium, Guaiacwood, Immortelle,

Lavender, Mandarin, Melissa, Mimosa, Myrtle, Neroli, Orange, Palmarosa, Patchouli, Peppermint, Petitgrain, Rose, Rose wood, Sandalwood, Vetiver, Ylang Ylang.

Contraindications: Sensitive skin, Pregnancy, Not for Babies, Excessive use can cause problems with bodily fluids such as phlegm, May have a narcotic effect and cause lack of concentration.

Properties: Menstrual regulator, Sedative, Antidepressant, Stimulates lactation, Boosts confidence, Analgesic, Balances hormones, Increases male fertility.

Uses: Stress, Skin Care, Throat infections, Coughs, Catarrh, Listlessness, Apathy, Melancholy, Hopelessness, Menstrual cramps, Back pain, Labour Pain, Impotence, Frigidity, Dry Skin, Itchy skin, Muscle aches and pains, Child Birth, Lethargy, Nervous tension, PMS, Post natal depression, Vaginal infections, Premature ejaculation, Hoarseness.

Juniper

Note - Middle

Type - Herb

Family - Cupressacea

Part - Berries

Extraction - Distillation

Aroma: Slightly woody

Blends well With: Benzoin, Bergamot, Cedarwood, Citrus oils, Cypress, Frankincense, Geranium, Ginger, Grapefruit, Orange, Lavender, Lemongrass, Lime, Melissa, Pine, Rosemary, Sandalwood, Vetiver.

Contraindications: Pregnancy, Kidney problems, not for use with babies.

Properties: Antiseptic, Anti-rheumatic, Antispasmodic, Aphrodisiac, Astringent, Carminative, Cicatrisant, Clears Mucous from intestines, Depurative, Detoxicant, Digestive aid, Disinfectant, Diuretic, Emmenagogue, Nervine, Insecticide, Parturient, Regulating appetite, Regulating periods, Rubefacient, Stimulating, Stomachic, Sudorific, Tonic, Vulnerary.

Uses: Genito-Urinary infections, Cystitis, Strangury (Inability to pass urine), Kidney stones, Cellulitis, Dropsy, Fluid retention, releasing toxins, Piles, Obesity, Cirrhosis, Arthritis, Rheumatism, Gout, Sciatica, Stiff joints, Menstrual cramps, Childbirth, Problem skin, Oily skin,

Congested skin, Seborrhoea, Acne, Blocked pores, Dermatitis, Eczema, Psoriasis, Swelling, Haemorrhoids, Ulcers, Stress, Oily Hair, Indigestion, Insomnia, Menstrual problems, Loss of appetite, Muscular aches and pains, PMS, Intestinal parasites, Colds, Flu, Anxiety, Purifying the blood, Circulation problems, Cough, Diarrhoea, Fatigue,

Lavender

Note - Middle

Type - Flower

Family - Labiatae

Part - Flowers

Extraction - Distillation

Aroma: Floral and slightly woody

Blends well With: Basil, Bay, Bergamot, Cedarwood, Chamomile, Citronella, Clary Sage, Cypress, Eucalyptus, Geranium, Grapefruit, Hyssop, Jasmine, Lemon, Lemongrass, Lime, Mandarin, Marjoram, Myrtle, Neroli, Nutmeg, Orange, Patchouli, Peppermint, Petitgrain, Pine, Thyme, Rosemary, Tea Tree.

Contraindications: Low Blood pressure, Early Pregnancy.

Properties: Analgesic, Anticonvulsive, Antidepressant, Antiphlogistic, Antirheumatic, Antiseptic, Antispasmodic, Antiviral, Bactericide, Balances central nervous system, Carminative, Cholagogue, Cicatrisant, Cordial, Cytophylactic, Decongestant, Deodorant, Detoxicant, Diuretic, Emmenagogue, Fungicide, Hypotensive, Increases gastric secretions, Nervine, Promotes growth of new skin cells, Restorative, Sedative, Splenetic, Stimulates bile production, Sudorific, Vulnerary.

Uses: Anger, Exhaustion, High blood pressure, Heart palpitations, Insomnia, Muscular spasm, Sprains, Strains, Rheumatic pains, Bronchitis, Asthma, Catarrh, Colds, Laryngitis, Throat infections, TB, Infections, Period pain, Scanty periods, Leucorrhoea, Childbirth, Cleanse spleen and liver, Nausea, Vomiting, Colic, Flatulence, Insect repellent, Burns, Sunburn, Acne, Eczema, Psoriasis, Abscesses, Boils, Carbuncles, Fungal growths, Swelling, Scars, Gangrene, Alopecia, Headaches, Lice, fleas, Appetite stimulating, Whooping cough, Flu, Nervous Tension, Stress, Vertigo, Shock, Athlete's foot, Genito-urinary problems, Cystitis, Scalds, Wounds, Sores, Varicose veins, PMT, Ulcers, Spider bites, Ringworm.

Lemon

Note - Top

Type - Citrus

Family - Rutaceae

Part - Peel

Extraction - Expression/Distillation

Aroma: Citrus, Sharp and fresh

Blends well With: Bay, Benzoin, Black pepper, Cardamon, Chamomile, Citronella, Citrus oils, Eucalyptus, Fennel, Frankincense, Geranium, Ginger, Jasmine, Juniper, Lavender, Linden, Neroli, Peppermint, Rose, Sandalwood, Ylang-Ylang.

Contraindications: Sensitive skin, Phototoxic, High Blood pressure.

Properties: Antacid, Antisclerotic, Antiscorbutic, Antineuralgic, Antirheumatic, Antipruritic, Antiseptic, Astringent, Bactericide, Carminative, Cicatrisant, Decongestant, Depurative, Diuretic, Emollient, Escharotic, Febrifuge, Haemostatic, Hepatic, Hypoglycemiant, Hypotensive, Insecticide, Laxative, Stomachic, Tonic, Vermifuge.

Uses: Varicose veins, Easing blood flow, High Blood pressure, Anaemia, Restoring vitality to Red blood cells, Stimulating white corpuscles, Infectious diseases, Nosebleeds, Sore throat, Coughs, Colds, Flu, Fever, Cold sores, Herpes, Digestive complaints, Counteracting acidity in stomach, Diabetes, Cleanses body, Constipation, Cellulite, Headaches, Migraine, Neuralgia, Rheumatism, Arthritis, Insect bites and stings, Removes dead skin cells, broken capillaries, Cleansing greasy skin, Corns, Warts, Verrucae, Scars, Strengthens nails, Cuts, Circulation problems, Boosts immune system, Asthma, Bronchitis, Catarrh, Dyspepsia, Sinusitis, Tonsillitis, Chilblains, Mouth sores, Middle ear infections, Depression, Indigestion, Boils, Debility, Fluid retention, Mouth Ulcers, Oily skin, Wounds.

Lemongrass

Note - Top

Type - Citrus

Family - Graminaea

Part - Leaves

Extraction - Distillation

Aroma: Strong, sweet and lemony

Blends well With: Basil, Cedarwood, Coriander, Eucalyptus, Geranium, Jasmine, Lavender, Neroli, Niaouli, Palmarosa, Peppermint, Rosemary, Tea Tree, Thyme, Vetiver, Yarrow.

Contraindications: Sensitive skin, use in low dose, High Blood pressure.

Properties: Antidepressant, Antiseptic, Bactericide, Carminative, Deodorant, Digestive, Diuretic, Fungicide, Galactagogue, Insecticide, Prophylactic, Stimulant, Tonic, Tones the skin, Sedative, Relaxing.

Uses: Exhaustion, Boosting parasympathetic nerves, Digestive complaints, Illness, Loss of appetite, Colitis, Indigestion, Gastro-enteritis, Infectious disease, Respiratory infections, Sore throat, Laryngitis, Fever, Aching muscles, Muscular aches and pains, Tired legs, Jet Lag, Headaches, Fatigue, Insect repellent, Fleas, Aids flow of breast milk, Open pores, Acne, Oily skin, Athlete's foot, Thrush, Digestive disturbances, Stress, Anxiety, Fluid retention, Gastric infections, Gout, Lice, Mental fatigue, Poor Circulation, Rheumatism, Scabies.

Lime

Note - Top

Type - Citrus

Family - Rutaceae

Part - Peel

Extraction - Expression/Distillation

Aroma: Sharp and Tangy

Blends well With: Angelica, Basil, Bergamot, Citronella, Citrus oils, Clary Sage, Geranium, Linden Blossom, Lavender, Neroli, Nutmeg, Palmarosa, Rose, Rosemary, Violet, Ylang Ylang.

Contraindications: Phototoxic, Sensitive skin.

Properties: Antiscorbic, Antiseptic, Antiviral, Apertif, Astringent, Bactericide, Disinfectant, Febrifuge, Haemostatic, Insecticide, Restorative, Tonic, Stimulating, Activating, Refreshing, Uplifting, Digestive stimulant, Stimulates appetite, Stimulates digestive secretions.

Uses: Apathy, Anxiety, Depression, Fever, Colds, Sore Throat, Flu, Coughs, Chest congestion, Catarrh, Sinusitis, Immune system tonic, Infections, Anorexia, Alcoholism, Rheumatic pain, Greasy skin,

Acne, Cuts, Wounds, Anaemia, Brittle nails, Boils, Chilblains, Corns, Herpes, Insect bites, Mouth Ulcers, Spots, Varicose veins, Warts, Arthritis, Cellulitis, High Blood Pressure, Nose bleeds, Obesity, Poor circulation, Rheumatism, Asthma, Throat infections, Bronchitis, Dyspepsia, Headaches, Stress.

Mandarin

Note - Top to Middle

Type - Citrus

Family - Rutacaea

Part - Peel

Extraction - Expression

Aroma: Sweet and tangy

Blends well With: Basil, Bergamot, Black Pepper, Chamomile, Cinnamon, Citrus oils, Clary Sage, Coriander, Cumin, Geranium, Grapefruit, Juniper, Lavender, Lemon, Lime, Marjoram, Nutmeg, Palmarosa, Petitgrain, Rose.

Contraindications: Phototoxic.

Properties: Antispasmodic, Cholagogue, Cytophylactic, Digestive tonic, Emollient, Sedative, Tonic, Uplifting, Stimulates appetite, Stimulates liver, Regulate Metabolic processes, Aids secretion of Bile, Aids breaking down of fats, Calms the Intestines, Revitalising, Skin tonic.

Uses: Depression, Anxiety, Flatulence, Morning sickness, PMT, Stretch Marks and scarring, Digestive weakness, Good for children/ pregnant women and the elderly, Insomnia, Nervousness, Acne, Oily skin, Fluid retention, Obesity, Hiccups, Restlessness, Constipation, Digestive problems.

Marjoram

Note - Middle

Type - Herb

Family - Labiatae

Part - Leaves

Extraction - Distillation

Aroma: Slightly Spicy

Blends well With: Basil, Bergamot, Cedarwood, Chamomile, Cypress, Eucalyptus, Geranium, Lavender, Mandarin, Melissa, Orange,

Nutmeg, Peppermint, Rosemary, Rose wood, Tea Tree, Thyme, Ylang Ylang.

Contraindications: May cause drowsiness with prolonged use, Pregnancy.

Properties: Analgesic, Anaphrodisiac, Antiseptic, Antispasmodic, Carminative, Cephalic, Cordial, Digestive, Emmenagogue, Expectorant, Hypotensive, Laxative, Nervine, Restorative, Sedative, Tonic, Vulnerary, Calming, Dilates the arteries and capillaries, Cleanses toxins, Regulates Menstrual cycle, Hypnotic,

Uses: Stress, Anxiety, Psychological trauma, Confronting "issues", Grief, Loneliness, Hyperactivity, Muscular aches and pains, Lower back pain, Rheumatic aches and pains, Swollen joints, Stiff joints, After-sports rub, High Blood pressure, Headache, Migraine, Insomnia, Digestive problems, Stomach cramps, Indigestion, Constipation, Flatulence, Sea Sickness, Chest infections, Colds, Sinusitis, Bronchitis, Asthma, Congestion, Painful periods, Bruises, Sprains, Arthritis, Menstrual problems, Coughs, Colic, Chilblains, Ticks, Lumbago, Strains, Dyspepsia, Leucorrhoea, PMS, Nervous tension, Cramps, Emotional comfort, Flu, Circulation.

Orange

Note - Top

Type - Citrus

Family - Citrus Vulgaris/Aurantium/ Sinesis

Part - Peel

Extraction - Expression

Aroma: Zesty and refreshing citrus

Blends well With: Angelica, Black Pepper, Cedarwood, Cinnamon, Citrus oils, Coriander, Clary Sage, Clove, Cumin, Cypress, Frankincense, Geranium, Ginger, Hyssop, Jasmine, Juniper, Lavender, Myrrh, Neroli, Nutmeg, Petitgrain, Rose, Rose wood.

Contraindications: Sensitive skin, Phototoxic.

Properties: Antidepressant, Antiseptic, Antispasmodic, Carminative, Digestive, Febrifuge, Sedative, Stomachic, Tonic, Stimulates bile, Helps digestion of fats, Stimulates Appetite, Aids digestion of Vitamin C, Aids formation of collagen, Lowers cholesterol, Uplifting, Calming.

Uses: Depression, Tension, Stress, Boredom, Lethargy, "Butterflies" in the stomach, Diarrhoea, Constipation, Viral infections,

Colds, Bronchitis, Fever, Repairing body tissues, Sore muscles, Rickety bones, Insomnia, Anxiety, Congested skin, Wrinkles, Dermatitis, Skin complaints, Flu, Oily skin, Coughs, Heartburn, Indigestion, Dull complexion, Dyspepsia, Mouth ulcers, Obesity, Palpitations, Water retention, Nervous tension.

Patchouli

Note - Base

Type - Herb

Family - Pogostemon patchouli

Part - Leaves

Extraction - Distillation

Aroma: Earthy and woody, a bit spicy (sort of mouldy)

Blends well With: Angelica, Bergamot, Black pepper, Cedarwood, Clary Sage, Elemi, Frankincense, Galbanum, Geranium, Ginger, Lavender, Lemongrass, Myrrh, Neroli, Orange Pine, Rose, Rose wood, Sandalwood, Ylang Ylang.

Contraindications: Sedative in low doses but Stimulating in high doses, May cause loss of appetite, May cause a headache in some people.

Properties: Grounding, Balancing, Antidepressant, Antipholgistic, Antiseptic, Anti-inflammatory, Aphrodisiac, Astringent, Cicatrisant, Cytophylactic, Deodorant, Diuretic, Febrifuge, Fungicide, Insecticide, Sedative, Tonic, Soothing, Calming, Opens pores, Cell regeneration.

Uses: Lethargy, Clarifying problems, Loose skin, Weight loss, Diarrhoea, Water Retention, Cellulite, Sweating, Insect bites and stings, Snake bites, Scars, Cracked skin, Sores, Wounds, Acne, Eczema, Fungal infections, Scalp disorders, Skin disorders, Athlete's foot, Dry Skin, Oily Skin, Tinea, Excessive Menstrual flow, Dandruff, Dermatitis, Hair care, Impetigo, Wrinkles, Frigidity, Nervous exhaustion, Stress, Anxiety, Bed sores, Depression.

Peppermint

Note - Top

Type - Herb

Family - Labiatae

Part - Leaves and flowering tops

Extraction - Distillation

Aroma: Minty

Blends well With: Basil, Benzoin, Cedarwood, Cypress, Eucalyptus, Jasmine, Lavender, Lemon, lemongrass, Mandarin, Marjoram, Niaouli, Pine, Rosemary.

Contraindications: (Very cooling on the skin - so best used in very small concentration on the skin), Sensitive skin, Pregnancy, Nursing Mothers (can discourage flow of milk), Should not be used with Homoeopathic remedies.

Properties: Analgesic, Antidontalgic, Anaesthetic, Antigalactagogue, Antiphlogistic, Antiseptic, Antispasmodic, Astringent, Carminative, Cephalic, Cholagogue, Cordial, Decongestant, Emmenagogue, Expectorant, Febrifuge, Hepatic, Nervine, Stimulant, Stomachic, Sudorific, Vasoconstrictor, Vermifuge, Encourages sweating, Insecticide, Softens skin, Improves thinking, Uplifting.

Uses: Anger, Hysteria, Nervous trembling, Mental fatigue, Depression, Colds, Mucous, Fever, Respiratory disorders, Dry coughs, Sinus congestion, Asthma, Bronchitis, Cholera, Pneumonia, Tuberculosis, Food poisoning, Vomiting, Diarrhoea, Constipation, Flatulence, Halitosis, Colic, Gall stones, Nausea, Travel sickness, Kidney disorders, Liver disorders, Numbness in the limbs, Shock, Vertigo, Anaemia, Dizziness, Fainting, Headaches, Migraines, Toothache, Aching feet, Rheumatism, Neuralgia, Muscular aches, Scanty menstruation, Painful periods, Mastitis, Dermatitis, Ringworm, Scabies, Pruritus, Itching, Inflammation, Sunburn, Blackheads, Oily hair and skin, Abdominal cramps, Digestive upsets, Flu, Morning Sickness, Shingles, Insect bites, Nervous stress, Indigestion, Acne, Palpitations.

Rosemary

Note - Middle

Type - Herb

Family - Labiatae

Part - Flowering tops and leaves

Extraction - Distillation

Aroma: A strong refreshing herbal scent

Blends well With: Basil, Black Pepper, Cedarwood, Frankincense, Geranium, Ginger, Grapefruit, Hyssop, Lavender, Lemongrass, Lime, Mandarin, Melissa, Myrtle, Orange, Peppermint, Petitgrain, Pine, Tea Tree, Tangerine.

Contraindications: Pregnancy, High Blood pressure, Epilepsy, Not to be taken with Homoeopathic remedies.

Properties: Analgesic, Antidepressant, Antirheumatic, Antiseptic, Antispasmodic, Astringent, Carminative, Cephalic, Cholagogue, Cicatrisant, Cordial, Decongestant, Digestive, Diuretic, Emmenagogue, Hepatic, Hypertensive, Nervine, Resolvent, Stimulant, Stomachic, Sudorific, Tonic, Vulnerary, Invigorating, Strengthens the mind, May help restore speech; hearing and sight, Heart tonic, Cardiac stimulant, Encourages hair growth.

Uses: Memory, Mental Strain, Dullness, Lethargy, Weakness, Mental exhaustion, Headaches, Migraines, Gastric problems, Vertigo, Paralysed limbs, Gout, Rheumatic pains, Tired and overworked muscles, Low blood pressure, Anaemia, Colds, Asthma, Chronic bronchitis, Flu, Hepatitis, Cirrhosis, Gallstones, Jaundice, Blocked bile ducts, Colitis, Dyspepsia, Flatulence, Stomach pains, Menstrual cramps, Scanty periods, Water retention, Cellulite, Obesity, Sagging skin, Skin congestion, Puffy skin, Swelling, Scalp disorders, Dandruff, Infections, Halitosis, Stress, Eyesight, Apathy, Muscle fatigue, Poor circulation, Aches and pains, Acne, Dermatitis, Eczema, Oily hair, Insect repellent, Lice, Seborrhoea, Scabies, Varicose veins, Rheumatism, Whooping cough.

Sandalwood

Note - Base

Type - Wood

Family - Santalaceae

Part - Inner wood

Extraction - Distillation

Aroma: Woody and Exotic

Blends well With: Basil, Benzoin, Bergamot, Black Pepper, Cedarwood, Cypress, Frankincense, Geranium, Jasmine, Lavender, Lemon, Mimosa, Myrrh, Neroli, Palmarosa, Patchouli, Rose, Rose wood, Vetiver, Ylang Ylang.

Contraindications: Depression? (May cause mood to drop lower, may act as an antidepressant)

Properties: Antiphlogistic, Antiseptic, Anti-inflammatory, Antispasmodic, Aphrodisiac, Astringent, Bechic, Carminative, Diuretic, Emollient, Expectorant, Sedative, Tonic, brings peace and acceptance, Stimulates immune system.

Uses: Nervous tension, Anxiety, Obsessional attitudes, Cutting ties with the past, Meditation, Genito-urinary problems, Cystitis,

Frigidity, Impotence, Chest infections, Sore throat, Dry coughs, Bronchitis, Lung infections, Insomnia, Catarrh, Heartburn, Diarrhoea, Dry skin, Oily skin, Aging skin, Dehydrated skin, Itching, Acne, Boils, Infected wounds, Cracked skin, Chapped skin, Shaving rash, PMS, Upset stomach, Stress, Urinary tract infections, Laryngitis, Nausea, Depression, Eczema, Fatigue, Respiratory problems, Skin problems, Sunstroke, Venereal infections.

Tea Tree

Note - Top

Type - Wood

Family - Myrtaceae

Part - Leaf

Extraction - Distillation

Aroma: Pungent and Sterile

Blends well With: Black Pepper, Cinnamon, Clary Sage, Clove, Coriander, Cumin, Cypress, Eucalyptus, Geranium, Ginger, Lavender, Lemon, Mandarin, Marjoram, Nutmeg, Orange, Pine, Rosemary, Thyme.

Contraindications: Sensitive skin, Pregnancy

Properties: Antibiotic, Antipuritic, Antiseptic, Antiviral, Antibacterial, Balsamic, Cicatrisant, Cordial, Disinfectant, Expectorant, Fungicide, Insecticide, Stimulant, Sudorific, Boosts immune system.

Uses: Shock, Infectious diseases, Eliminating toxins, Flu, Colds, Cold sores, Catarrh, Glandular fever, Gingivitis, AIDS (not a cure, but may boost immune system to be of benefit), Post-operative shock, Convalescence, Vaginal thrush, Genital infections, Urinary tract infections, Cystitis, Genital and anal pruritus, Chickenpox, Itching, Rashes, Insect bites and stings, Ear infections, Tonsillitis, Enteritis, Intestinal parasites, Infected wounds, Boils, Carbuncles, Spots, Acne, Shingles, Burns, Sores, Sunburn, Ringworm, Warts, Tinea, Herpes, Athlete's foot, Dry scalp, Dandruff,. Bronchitis, Verrucae, Asthma, Hysteria, Abscesses, Calluses, Blisters, Respiratory problems.

6

Types of Aromatic Plants

Aromatic Fire-cured smoking tobacco is a robust variety of tobacco used as a condimental for pipe blends. It is cured by smoking over gentle fires. In the United States, it is grown in the western part of Tennessee, Western Kentucky and in Virginia. Fire-cured tobacco grown in Kentucky and Tennessee is used in some chewing tobaccos, moist snuff, some cigarettes and as a condiment leaf in pipe tobacco blends. It has a rich, slightly floral taste, and adds body and aroma to the blend.

Another fire-cured tobacco is "Latakia" and is produced from oriental varieties of *N. tabacum*. The leaves are cured and smoked over smoldering fires of local hardwoods and aromatic shrubs in Cyprus and Syria. Latakia has a pronounced flavour and a very distinctive smoky aroma, and is used in Balkan and English-style pipe tobacco blends.

Brightleaf Tobacco

Brightleaf is commonly known as "Virginia tobacco", often regardless of which state they are planted. Prior to the American Civil War, most tobacco grown in the US was fire-cured dark-leaf. This type of tobacco was planted in fertile lowlands, used a robust variety of leaf, and was either fire cured or air cured.

Sometime after the War of 1812, demand for a milder, lighter, more aromatic tobacco arose. Ohio, Pennsylvania and Maryland all innovated quite a bit with milder varieties of the tobacco plant. Farmers around the country experimented with different curing processes. But the breakthrough didn't come until around 1839. Brightleaf tobacco leaf ready for harvest. When it turns yellow-green the sugar content is at its peak, and it will cure to a deep golden colour with mild taste.The leaves are harvested progressively up the stem from the base, as they ripen. It had been noticed for centuries that sandy, highland soil produced thinner, weaker plants.

Captain Abisha Slade, of Caswell County, North Carolina had a good deal of infertile, sandy soil, and planted the new "gold-leaf" varieties on it. Slade owned a slave, Stephen, who around 1839 accidentally produced the first real bright tobacco. He used charcoal to restart a fire used to cure the crop. The surge of heat turned the leaves yellow. Using that discovery, Slade developed a system for producing bright tobacco, cultivating on poorer soils and using charcoal for heat-curing.

Slade made many public appearances to share the bright-leaf process with other farmers. Prosperous and outgoing, he built a brick house in Yanceyville, North Carolina, and at one time had many servants. News spread through the area pretty quickly. The infertile sandy soil of the Appalachian piedmont was suddenly profitable, and people rapidly developed flue-curing techniques, a more efficient way of smoke-free curing. Farmers discovered that Bright leaf tobacco needs thin, starved soil, and those who could not grow other crops found that they could grow tobacco. Formerly unproductive farms reached 20-35 times their previous worth. By 1855, six Piedmont counties adjoining Virginia ruled the tobacco market.

By the outbreak of the Civil War, the town of Danville, Virginia actually had developed a bright-leaf market for the surrounding area in Caswell County, North Carolina and Pittsylvania County, Virginia. Danville was also the main railway head for Confederate soldiers going to the front. These brought bright tobacco with them from Danville to the lines, traded it with each other and Union soldiers, and developed quite a taste for it. At the end of the war, the soldiers went home and suddenly there was a national market for the local crop. Caswell and Pittsylvania counties were the only two counties in the South that experienced an *increase* in total wealth after the war.

White Burley

In 1864, George Webb of Brown County, Ohio planted Red Burley seeds he had purchased, and found that a few of the seedlings had a whitish, sickly look. He transplanted them to the fields anyway, where they grew into mature plants but retained their light colour. The cured leaves had an exceedingly fine texture and were exhibited as a curiosity at the market in Cincinnati.

The following year he planted ten acres (40,000 m^2) from seeds from those plants, which brought a premium at auction. The air-cured leaf was found to be mild tasting and more absorbent than any other variety. *White Burley*, as it was later called, became the main component in chewing tobacco, American blend pipe tobacco, and American-style cigarettes. The white part of the name is seldom used today, since red burley, a dark air-cured variety of the mid-1800s, no longer exists.

Shade Tobacco

It is not well known that the northern US states of Connecticut and Massachusetts are also one of the important tobacco-growing regions of the country. Long before Europeans arrived in the area, Native Americans harvested wild tobacco plants that grew along the banks of the Connecticut River. Today, the Connecticut River valley north of Hartford, Connecticut is known as Tobacco Valley, and the fields and drying sheds are visible to travelers on the road to and from Bradley Field, the major Connecticut airport. The tobacco grown here is known as shade tobacco, and is used as outer wrappers for some of the world's finest cigars.

Early Connecticut colonists acquired from the Native Americans the habit of smoking tobacco in pipes and began cultivating the plant commercially, even though the Puritans referred to it as the "evil weed". The plant was outlawed in Connecticut in 1650, but in the 1800s as cigar smoking began to be popular, tobacco farming became a major industry, employing farmers, laborers, local youths, southern African Americans, and migrant workers.

Working conditions varied from pleasant summer work for students, to backbreaking exploitation of migrants. Each tobacco plant yields only 18 leaves useful as cigar wrappers, and each leaf requires a great deal of individual manual attention during harvesting. Although temperatures in the curing sheds sometimes exceeds 100 degrees F., no work is done inside the sheds while the tobacco is being fired.

In 1921, Connecticut tobacco production peaked, at 31,000 acres (125 km^2) under cultivation. The rise of cigarette smoking and the decline of cigar smoking has caused a corresponding decline in the demand for shade tobacco, reaching a minimum in 1992 of 2,000 acres (8 km^2) under cultivation. Since then, however, cigar smoking has become more popular again, and in 1997 tobacco farming had risen to 4,000 acres (16 km^2).

However, only 1,050 acres of shade tobacco were harvested in the Connecticut Valley in 2006. Connecticut seed is being grown in Ecuador, where labour is very cheap. The industry has weathered some major catastrophes, including a devastating hailstorm in 1929, and an epidemic of brown spot fungus in 2000, but is now in danger of disappearing altogether, given the value of the land to real estate speculators. The older and much less labour intensive Broadleaf plant, which produces an excellent maduro wrapper as well as binder and filler for cigars, is increasing in acreage in the Connecticut Valley.

Perique

Perhaps the most strongly-flavoured of all tobaccos is the Perique, from Saint James Parish, Louisiana. When the Acadians made their way into this region in 1755, the Choctaw and Chickasaw tribes were cultivating a variety of tobacco with a distinctive flavour. A farmer called Pierre Chenet is credited with first turning this local tobacco into the Perique in 1824 through the technique of pressure-fermentation.

The tobacco plants are manually kept suckerless, and pruned to exactly 12 leaves, through their early growth. In late November, when the leaves are a dark, rich green and the plants are 23-30 inches (600 to 751 mm) tall, the whole plant is harvested in the late evening and hung to dry in a sideless curing barn. Once the leaves have partially dried, but while still supple (usually less than 2 weeks in the barn), any remaining dirt is removed and the leaves are moistened with water and stemmed by hand. The leaves are then rolled into "torquettes" of approximately 1 pound (449 g) and packed into hickory whiskey barrels.

The tobacco is then kept under pressure using oak blocks and massive screw jacks, forcing nearly all the air out of the still-moist leaves. Approximately once a month, the pressure is released, and each of the torquettes is "worked" by hand to permit a little air back into the tobacco. After a year of this treatment, the Perique is ready for consumption, although it may be kept fresh under pressure for many years. Extended exposure to air degrades the particular character of the Perique. The finished tobacco is dark brown, nearly black, very moist with a fruity, slightly vinegary aroma.

Considered the truffle of pipe tobaccos, the Perique is used as a component of many blended pipe tobaccos, but is too strong to be smoked pure. At one time, the freshly moist Perique was also chewed, but none is now sold for this purpose. Less than 16 acres (65,000 m^2) of this crop remain in cultivation, most by a single farmer called Percy Martin, in Grande Pointe, Louisiana. For reasons unknown, the particular flavour and character of the Perique can only be acquired on a small triangle of Saint James Parish, less than 4 by 10 miles

(5 by 16 km). Although at its peak, Saint James Parish was producing around 20 tons of the Perique a year, output is now merely a few barrelsful. It is traditionally a pipe tobacco, and is still very popular with pipe-smokers, typically blended with pure Virginia to lend spice, strength, and coolness to the blend. Perique may now also be found in the Perique cigarettes of Santa Fe Natural Tobacco Co., in an approximately 1 part to 5 blend with lighter tobaccos. A similar tobacco, based on pressure-fermented Kentucky tobacco is available by the name Acadian Yellow River Perique.

Oriental Tobacco

Oriental tobacco is a sun-cured, highly aromatic, small-leafed variety that is grown in Turkey, Greece, Bulgaria, and Macedonia. Oriental tobacco is frequently referred to as "Turkish tobacco", as these regions were all historically part of the Ottoman Empire. Many of the early brands of cigarettes were made mostly or entirely of Oriental tobacco; today, its main use is in blends of pipe and especially cigarette tobacco (a typical American cigarette is a blend of bright Virginia, burley and Oriental).

Tobacco Products

Snuff

Snuff is a generic term for fine-ground smokeless tobacco products. Originally the term referred only to dry snuff, a fine tan dust popular mainly in the eighteenth century. This is often called "Scotch Snuff", a folk-etymology derivation of the scorching process used to dry the cured tobacco by the factory. Snuff powder originated in the UK town of Great Harwood and was famously ground in the town's monument prior to local distribution and transport further up north to Scotland.

Snuff has been found to be beneficial in some cases of hay fever due to the fact that the snuff may prevent allergens from getting to the mucus membrane within the nose. Most would say, however, that the detrimental effects of Nicotine outweigh the benefits.

Types of Snuff

European (*dry*) snuff is intended to be *sniffed* up the nose. Snuff is not "snorted" because snuff shouldn't get past the nose, *i.e.*; into sinuses, throat or lungs. European snuff comes in several varieties: Plain, Toast (fine ground - *very* dry), "Medicated" (menthol, camphor, eucalyptus, etc.), Scented, and Schmalzler, a German variety. The major brand names of European snus are: Bernards (Germany),

Fribourg & Treyer (UK), Gawith (UK), Gawith Hoggarth (UK), Hedges (UK), Lotzbeck (Germany), McChrystal's (UK), Pöschl (Germany) and Wilsons of Sharrow (UK), TUTUN-CTC (Moldova).

American snuff is much stronger, and is intended to be dipped. It comes in two varieties — "sweet" and "salty". Until the early 20th century, snuff dipping was popular in the United States among rural people, who would often use sweet barkless twigs to apply it to their gums. Popular brands are Tube Rose and Navy.

The second, and more popular in North America, variety of snuff is moist snuff, or dipping tobacco (sometimes known as *smokeless tobacco*). This practice is known as "dipping." In the Southern states, taking a "dip" of moist snuff is called "putting a rub in," the moist snuff in the mouth is known as a "rub." This is occasionally referred to as "snoose" in New England and the Midwest and is derived from the Scandinavian word for snuff, "snuff". Like the word, the origins of moist snuff are Scandinavian, and the oldest American brands indicate that by their names. Snuff is also called a "ding" in New England (*i.e.* "Packing a ding"). American Moist snuff is made from dark fire-cured tobacco that is ground, sweetened, and aged by the factory. Prominent North American brands are Copenhagen, Skoal, Timber Wolf, Chisholm, Grizzly, and Kodiak. American moist snus tends to be dipped. Some modern *smokeless tobacco* brands, such as Kodiak, have an aggressive nicotine delivery. This is accomplished with a higher dose of nicotine than cigarettes, a high pH level (which helps nicotine enter the blood stream faster), and a high portion of unprotonated (free base) nicotine.

It has been suggested by *The Economist* magazine that the ban on smoking tobacco indoors in some areas, such as Britain and New York City, may lead to a resurgence in the popularity of snuff as an alternative to tobacco smoking. Although the large-scale closure of British mines in the 1980s deprived the snuff industry of its major market since snuff became unfashionable (miners took snuff underground instead of smoking to avoid lethal explosions and fires), sales at Britain's largest snuff retailer have reportedly been rising at about 5% per year. Snuff is tobacco that you can snort

Chewing Tobacco

Chewing is one of the oldest ways of consuming tobacco leaves. Native Americans in both North and South America chewed the leaves of the plant, frequently mixed with lime. Modern chewing tobacco is produced in three forms: twist, plug, and scrap. A few manufacturers in the United Kingdom produce particularly strong twist tobacco meant for use in smoking pipes rather than chewing. These twists are not mixed with lime although they may be flavoured with whiskey, rum, cherry or other flavors common to pipe tobacco.

Twist is the oldest form. One to three high-quality leaves are braided and twisted into a rope while green, and then are cured in the same manner as other tobacco.

Originally devised by sailors due to fire hazards of smoking at sea; and until recently this was done by farmers for their personal consumption in addition to other tobacco intended for sale. Modern twist is occasionally lightly sweetened. It is still sold commercially, but rarely seen outside of Appalachia. Popular brands are Mammoth Cave, Moore's Red Leaf, and Cumberland Gap. Users cut a piece off the twist and chew it, expectorating.

Plug chewing tobacco is made by pressing together cured tobacco leaves in a sweet (often molasses-based) syrup. Originally this was done by hand, but since the second half of the 19th century leaves were pressed between large tin sheets. The resulting sheet of tobacco is cut into plugs. Like twist, consumers sometimes cut, but more often bite off a piece of the plug to chew. Major brands are Days O Work and Cannonball. Scrap, or looseleaf chewing tobacco, was originally the excess of plug manufacturing. It is sweetened like plug tobacco, but sold loose in bags rather than a plug. Looseleaf is one of the more popular forms of tobacco in modern times. Among those, popular brands are Red Man, Beechnut, Mail Pouch and Southern Pride. Looseleaf chewing tobacco can also be dipped. Of course, as far as people go, a person could prefer any type of chewing tobacco.

Snus

Swedish snus is different in that it is made from steam-cured tobacco, rather than fire-cured, and its health effects are markedly different, with epidemiological studies showing dramatically lower rates of cancer and other tobacco-related health problems than cigarettes, American "Chewing Tobacco", Indian Gutka or African varieties. Prominent Swedish brands are Swedish Match, Ettan, and Tree Ankare. In the Scandinavian countries, moist snuff comes either in loose powder form, to be pressed into a small ball or ovoid either by hand or with the use of a special tool. It is sometimes packaged in small bags, suitable for placing inside the upper lip, called "portion snuff".

Figure: *"White portion" snus of the Swedish label General.*

In the United States, the Skoal brand of moist snuff distributes a similar product, packed with standard American moist snuff, often flavoured with fruits or liquors; these small bags are called "Skoal Bandits." These small bags keep the loose tobacco from becoming lodged between the user's teeth; they also generate less spittle when in contact with mucous membranes inside the mouth which extends the usage time of the tobacco product.

Since it is not smoked, snuff in general generates less of the nitrosamines and other carcinogens in the tar that forms from the partially anaerobic reactions in the smoldering smoked tobacco. The steam curing of snus rather than fire-curing or flue-curing of other smokeless tobaccos has been demonstrated to generate even fewer of such compounds than other varieties of snuff; 2.8 parts per mil for *Ettan* brand compared to as high as 127.9 parts per mil in American brands, according to a study by the State of Massachusetts Health Department. It is hypothesized that the widespread use of snus by Swedish men (estimated at 30% of Swedish men, possibly because it is much cheaper than cigarettes), displacing tobacco smoking and other varieties of snuff, is responsible for the incidence of tobacco-related mortality in men being significantly lower in Sweden than any other European country. In contrast, since women are much less likely to use snus, their rate of tobacco-related deaths in Sweden is similar to that in other European countries. Snus is clearly less harmful than other tobacco products; according to Kenneth Warner, director of the University of Michigan Tobacco Research Network,

"The Swedish government has studied this stuff to death, and to date, there is no compelling evidence that it has any adverse health consequences. ... Whatever they eventually find out, it is dramatically less dangerous than smoking."

Public health researchers maintain that, nevertheless, even the low nitrosamine levels in snus cannot be completely risk free, but snus proponents maintain that inasmuch as snus is used as a substitute for smoking or a means to quit smoking, the net overall effect is positive, similar to the effect of nicotine patches, for instance. Snus is banned in the European Union countries outside of Sweden (regular snus, not portion, is allowed in Denmark and snus is also becoming a regular among Norwegians, as cigarettes are seen by Norwegian popular culture as untrendy and much more unhealthy than snus). Although this is officially for health reasons, it is widely regarded, in fact, as being for economic reasons, since other smokeless tobacco products (mainly from India) associated with much greater risk to health are sold too.

Although it lacks the carcinogenicity of high levels of nitrosamines, however, any harmful effects of nicotine will still be seen with snus usage. Current research concentrates on nicotine's effect on the circulatory system and on the pancreas. On June 11, 2006, Reynolds Tobacco announced that it would be test marketing Camel brand snus in Portland, Oregon and Austin, Texas by the end of the month. The product would be manufactured in Sweden, in conjunction with British American Tobacco, manufacturers of BAT snus.

Creamy Snuff

Creamy snuff is a tobacco paste, consisting of tobacco, clove oil, glycerin, spearmint, menthol, and camphor, and sold in a toothpaste tube. It is marketed mainly to women in India, and is known by the brand names Ipco (made by Asha Industries), Denobac, Tona, Ganesh. It is locally known as "mishri" in some parts of Maharashtra. According to the U.S NIH-sponsored 2002 Smokeless Tobacco Fact Sheet, it is marketed as a dentifrice. The same factsheet also mentions that it is "often used to clean teeth". The manufacturer recommends letting the paste linger in the mouth before rinsing.

Tobacco Water

Tobacco water is a traditional organic insecticide used in domestic gardening. Tobacco dust can be used similarly.

It is produced by boiling strong tobacco in water, or by steeping the tobacco in water for a longer period. When cooled the mixture can be applied as a spray, or 'painted' on to the leaves of garden plants, where it will prove deadly to insects. Basque *angulero* fishermen kill immature eels (elvers) in an infusion of tobacco leaves before parboiling them in salty water for transportation to market as *angulas*, a seasonal delicacy.

Campaigns

There have been many campaigners against the growth of tobacco. New Zealand companies Reuben and Co. and Joseph's Nachos continue to protest against the growth and sale of tobacco throughout the world. Other well known protesters include Lance Tollenaar and Tim Smith, as well as the company owners, Joseph Griffiths and Reuben Palmer.

Sugarcane

Sugarcane or Sugar cane (*Saccharum*) is a genus of 6 to 37 species (depending on taxonomic interpretation) of tall grasses (family Poaceae, tribe Andropogoneae), native to warm temperate to tropical regions of the Old World. They have stout, jointed, fibrous stalks that are rich in sugar and measure 2 to 6 meters tall. All of the sugarcane species interbreed, and the major commercial cultivars are complex hybrids.

Figure: *Cut sugar cane*

Cultivation and Uses

About 200 countries grow the crop to produce 1,324 million tons (more than six times the amount of sugar beet produced). As of the year 2005, the world's largest producer of sugar cane by far is Brazil. Uses of sugar cane include the production of sugar, Falernum, molasses, rum, cachaça (the national spirit of Brazil) and ethanol for fuel. The bagasse that remains after sugarcane crushing is used to provide both heat energy, used in the mill, and electricity, which is typically on-sold to the consumer electricity grid.

History

Sugarcane was originally from tropical Southeast Asia. Different species likely originated in different locations with *S. barberi* originating in India and *S. edule* and *S. officinarum* coming from New Guinea. The thick stalk stores energy as sucrose in the sap. From this juice, sugar is extracted by evaporating the water. Crystallized sugar was reported 2500 years ago in India. Around the eighth century A.D., Arabs introduced sugar to the Mediterranean and it was cultivated in Spain. It was among the early crops brought to the Americas by Spaniards. Sugarcane was, and is still is, extensively grown in the Caribbean, where it was first brought by Christopher Columbus during his second voyage to The Americas, initially to the island of Hispaniola. In colonial times, sugar was a major product of the triangular trade of New World raw materials, European manufactures, and African slaves. France found its sugarcane islands so valuable it effectively traded Canada to Britain for their return of Guadeloupe, Martinique and St. Lucia at the end of the Seven Years' War.

The Dutch similarly kept Suriname, a sugar colony in South America, instead of seeking the return of the New Netherlands (New Amsterdam). Cuban sugarcane produced sugar that received price supports from and a guaranteed market in the USSR; the dissolution of that country forced the closure of most of Cuba's sugar industry. Sugarcane remains an important part of the economy of Belize, Barbados, the Dominican Republic, Guadeloupe, Jamaica, Grenada, and other islands. The sugarcane industry is a major export for the Caribbean, but it is expected to collapse with the removal of European preferences by 2009.

Sugarcane production greatly influenced many tropical Pacific islands, most particularly Hawaii and Fiji. In these islands, sugar came to dominate the economic and political landscape after the indigenous societies had been invaded by Europeans and Americans, who promoted immigration from various Asian countries for workers to tend and harvest the crop. Sugar-industry policies eventually established the ethnic makeup of the island populations that now exist, profoundly affecting modern politics and society in the islands.

Brazil is a major grower of sugarcane, which is used to produce sugar and provide the ethanol used in making gasoline-ethanol blends (gasohol) for transportation fuel.

Cultivation

Sugarcane cultivation requires a tropical or subtropical climate, with a minimum of 600 mm (24 in) of annual moisture. It is one of the most efficient photosynthesizers in the plant kingdom, able to convert up to 2 percent of incident solar energy into biomass. In prime growing regions, such as Hawaii, sugarcane can produce 20 kg for each square meter exposed to the sun. Sugarcane is propagated from cuttings, rather than from seeds; although certain types still produce seeds, modern methods of stem cuttings have become the most common method of reproduction. Each cutting must contain at least one bud, and the cuttings are usually planted by hand. Once planted, a stand of cane can be harvested several times; after each harvest, the cane sends up new stalks, called ratoons. Usually, each successive harvest gives a smaller yield, and eventually the declining yields justify replanting. Depending on agricultural practice, two to ten harvests may be possible between plantings.

Sugarcane is harvested by hand or mechanically. Hand harvesting accounts for more than half of the world's production, and is especially dominant in the developing world. When harvested by hand, the field is first set on fire. The fire spreads rapidly, burning away dry dead leaves, and killing any venomous snakes hiding in the crop, but leaving the water-rich stalks and roots unharmed. With knives (usually Cane Knives, but Machetes are also commonly used), harvesters then cut the standing cane just above the ground. A skilled harvester can cut 500 kg of sugarcane in an hour. With mechanical harvesting, a sugarcane combine (or chopper harvester), a harvesting machine originally developed in Australia, is used. The machine cuts the cane at the base of the stalk, separates the cane from its leaves, and deposits the cane into a cart while blowing the cut leaves back onto the field. Such machines can harvest 30 tonnes of cane each hour, but cane harvested using these machines must be transported to the processing plant rapidly; once cut, sugarcane begins to lose its sugar content, and damage inflicted on the cane during mechanical harvesting accelerates this decay.

Pests

The most important sugarcane pests are the larvae of some lepidoptera species, including turnip moth, the sugarcane borer, *Diatraea saccharalis* and the Mexican rice borer (*Eoreuma loftini*), leaf-cutting ants, termites, spittlebugs (especially *Mahanarva fimbriolata* and *Deois flavopicta*) and the beetle *Migdolus fryanus*, among others.

Diseases

Processing

Traditionally, sugarcane has been processed in two stages. Sugarcane mills, located in sugarcane-producing regions, extract sugar from freshly harvested sugarcane, resulting in raw sugar for later refining, and in "mill white" sugar for local consumption.

Sugar refineries, often located in heavy sugar-consuming regions, such as North America, Europe, and Japan, then purify raw sugar to produce refined white sugar, a product that is more than 99 percent pure sucrose. These two stages are slowly becoming blurred. Increasing affluence in the sugar-producing tropics has led to an increase in demand for refined sugar products in those areas, where a trend toward combined milling and refining has developed.

Milling

In a sugar mill, sugarcane is washed, chopped, and shredded by revolving knives. The shredded cane is repeatedly mixed with water and crushed between rollers; the collected juices (called garapa in Brazil) contain 10–15 percent sucrose, and the remaining fibrous solids, called bagasse, are burned for fuel. Bagasse makes a sugar mill more than self-sufficient in energy; the surplus bagasse can be used for animal feed, in paper manufacture, or burned to generate electricity for the local power grid.

The cane juice is next mixed with lime to adjust its pH to 7. This mixing arrests sucrose's decay into glucose and fructose, and precipitates out some impurities. The mixture then sits, allowing the lime and other suspended solids to settle out, and the clarified juice is concentrated in a multiple-effect evaporator to make a syrup about 60 percent by weight in sucrose.

This syrup is further concentrated under vacuum until it becomes supersaturated, and then seeded with crystalline sugar. Upon cooling, sugar crystallizes out of the syrup. A centrifuge is used to separate the sugar from the remaining liquid, or molasses. Additional crystallizations may be performed to extract more sugar from the molasses; the molasses remaining after no more sugar can be extracted from it in a cost-effective fashion is called blackstrap. Raw sugar has a yellow to brown colour. If a white product is desired, sulphur dioxide may be bubbled through the cane juice before evaporation; this chemical bleaches many colour-forming impurities into colourless ones. Sugar bleached white by this *sulfitation* process is called "mill white," "plantation white," and "crystal sugar." This form of sugar is the form most commonly consumed in sugarcane-producing countries.

Refining

In sugar refining, raw sugar is further purified. It is first mixed with heavy syrup and then centrifuged clean. This process is called "affination"; its purpose is to wash away the outer coating of the raw sugar crystals, which is less pure than the crystal interior. The remaining sugar is then dissolved to make a syrup, about 70 percent by weight solids.

The sugar solution is clarified by the addition of phosphoric acid and calcium hydroxide, which combine to precipitate calcium phosphate. The calcium phosphate particles entrap some impurities and absorb others, and then float to the top of the tank, where they can be skimmed off. An alternative to this "phosphatation" technique is "carbonatation," which is similar, but uses carbon dioxide and calcium hydroxide to produce a calcium carbonate precipitate.

After any remaining solids are filtered out, the clarified syrup is decolorized by filtration through a bed of activated carbon; bone char was traditionally used in this role, but its use is no longer common. Some remaining colour-forming impurities adsorb to the carbon bed.

The purified syrup is then concentrated to supersaturation and repeatedly crystallized under vacuum, to produce white refined sugar. As in a sugar mill, the sugar crystals are separated from the molasses by centrifugation. Additional sugar is recovered by blending the remaining syrup with the washings from affination and again crystallizing to produce brown sugar. When no more sugar can be economically recovered, the final molasses still contains 20–30 percent sucrose and 15–25 percent glucose and fructose. To produce granulated sugar, in which the individual sugar grains do not clump together, sugar must be dried. Drying is accomplished first by drying the sugar in a hot rotary dryer, and then by conditioning the sugar by blowing cool air through it for several days.

Ribbon Cane Syrup

Ribbon cane is a subtropical type that was once widely grown in southern United States, as far north as coastal North Carolina. The juice was extracted with horse or mule-powered crushers; the juice was boiled, like maple syrup, in a flat pan, and then used in the syrup to form as a sweetener for other foods. It is not a commercial crop nowadays, but a few growers try to keep alive the old traditions and find ready sales for their product. Most sugarcane production in the United States occurs in Florida and Louisiana, and to a lesser extent in Hawaii and Texas.

Sugarcane as Food

In most countries where sugarcane is cultivated, there are several foods and popular dishes derived from it, such as:

- Direct consumption of raw sugarcane cylinders or cubes, which are chewed to extract the juice, and the bagasse is spat out
- Freshly extracted juice (garapa, *guarab, guarapa, guarapo, papelón,* or *caldo de cana*) by hand or electrically operated small mills, with a touch of lemon and ice, makes a delicious and popular drink.
- Molasses, used as a sweetener and as a syrup accompanying other foods, such as cheese or cookies
- Rapadura, a candy made of flavoured solid brown sugar in Brazil, which can be consumed in small hard blocks, or in pulverized form (flour), as an add-on to other desserts.
- Sugarcane is also used in rum production, especially in the Caribbean.

Propagation-Fruit Plants

In case of fruits like Pomelo, Limes, etc. they bear fruits in 3 to 4 years, Litchi and mango take six to eight years before a bumper crop can be gathered. The plants raised by vegetative propagation will come to flowering and fruiting much earlier than the one obtained from seed.

They might bear at the very first year after planting, but precocious bearing should not be encouraged. It will hamper the vigour of the fruit tree. Where seedlings are planted, the fruiting period is protracted and one is never certain of obtaining a good type from a seedling unless careful selection has been practised. Plants propagated by vegetative means retain the good qualities of their mother without any deterioration.

For instance, a seedling Kaghzi lime of a thin skinned mother, may have a thick rind and little juice while one propagated by gooties will perpetuate the thin skin character. Always buy from a reliable source as foliage differences cannot be depended on for judging the quality and type of the plant. Because to discover after eight years that a Langra Mango tree bears small acid fruits will be no recompense from having purchased the graft a few paisa cheaper from an unauthentic source.

Major Fruits

- *Banana (Musa paradisiaca)*: For vegetative propagation sucker is the most popular planting material. Among the suckers which arise from the rhizome, the sword suckers (having narrow sword-shaped leaves) are better than the water suckers (with broad leaves) and should be used.
- *Cashew (Anacardium occidentale):* The plants are propagated by seeds, air-layering, inarching and side grafting.
- *Oranges:* Oranges in India can be mainly divided into two groups—Sweet oranges (*Citrus sinensis*) The typical examples of sweet oranges in our country are Malta, Mosambi, Sathgudi and Washington Navel, etc., and of Mandarine group are all types of loose skinned oranges commonly known as Nagpur Santra, Assam Santra, Coorg Santra and Sikkim orange etc. Unlike the Mandarine orange, sweet oranges are tight skinned and heavy.

Sweet Oranges: Sweet oranges are usually propagated by budding.

Mandarine Oranges (C. reticulata): Budding on rough lemon and seed propagation is better prevalent in Santra oranges.

- *Lime (C. aurantifolia):* Seed propagation and budding are usually practised in sour lime. Sweet lime is propagated both by stem cutting and seed.
- *Shaddock (C. decumana):* Plants should be grown from gooties or buddlings.
- *Custard apple (Anona squoamosa):* Seedlings vary appreciably in their characters. Grafting by budding and inarching are usually characters. Grafting by budding and inarching are usually done on Bullock's Heart (*Anona reticulata*) plants, which is an allied species.
- *Grape (Vitis vinigera):* The grapes are usually propagated by stem cuttings. Seed propagation is used fro breeding purposes.
- *Guava (Psidium guajava):* It is propagated by seed as well as by air-layering, ground layering and inarching.
- *Jack fruit (Artocarpus integrifolia):* It is propagated by seed and inarching.

- *Litchi (Litchi chinensis):* Air layering is the usual method of propagation. Ground layering, budding, grafting and inarching are also practised. Plants, which are propagated by gooties, bear fruits in 4-5 years.
- *Mango (Mangifera indica):* Mango can be propagated by seed and by vegetative means such as budding, grafting, inarching, etc. Vegetatively propagated plants usually give good quality fruits. They also bear much earlier than the seed propagated plants.
- *Papaya (Carica papaya):* Papaya is propagated by seed, which should be sown at the early monsoon. Seedlings of 30 to 45 days age are 20-30 cms. Height should be transplanted in monsoon 2.5 to 4 metres apart.
- *Pineapple (Ananas comosus):* Pineapple grows from suckers, stumps, side shoots or crowns, but is usually propagated from slips and suckers. Pineapple suckers will give fruit a year ahead of plants raised from fruit crowns.
- *Bael (Aegle marmelos):* The plants are usually propagated by seed. Seedlings bear in 6-8 years but give best result after the 10th year.
- *Ber (Zizyphus jujuba):* It is usually seed propagated but shield and ring budding are also common.
- *Loquat (Eribotrya japonica):* The plants are propagated by gooties, which bear in 2-4 years. Seedlings take about 8 to 10 years to fruit.
- *Pomegranate (Punica granatum):* Plants raised through cuttings or gooties bear in 3 years while seedlings take 6-8 years and are never to be depended on.
- *Sapota (Achras sapota):* Ground layering, air layering and inarching are the usual methods of propagation.

Propagation of Ornamental Plants

Plants usually reproduce in two ways:

1. By seeds and
2. By vegetative parts of plant.

The letter method is very popular in the multiplication of fruit and ornamental plants.

Sexual Propagation or Propagation by Seeds: This is the easiest method of propagation of plants. In this method the seeds are sown, covered with a layer of soil or leaf mould and watered. After germination the seedlings are allowed to grow up to 4-leaf stage and then they are transplanted in the beds or pots. In some cases the seeds are directly sown in the ground. These seeds originate by the union of male and female gamete.

During pollination pollens of male reproductive organ female reproductive organ. When the ovules are fertilized by the pollens of the same flower the process is known as self-pollination. Cross-pollination occurs when the pollens come from a different source and usually the pollinating agents are wind or insects. Self-pollination occurs when the pollens come from a different source and usually the pollinating agents are wind or insects.

Self-pollinated seeds are likely true to type or variety, but the cross-pollinated ones may not resemble the parents for all the character. After pollination, male gametes fertilize ovules resulting in the production of seed.

A Sexual or Vegetative Propagation: The following are the reasons for propagating plants vegetatively:

1. Many plants do not produce seeds under local condition or have lost the ability of production of viable seed. It is found in several cases that plants which readily root from cuttings do not produce seeds *e.g.*, *Acalypha*, *Eranthemum*.
2. Plants, which are cross-pollinated and have different varieties in cultivation, produce seeds of heterogeneous mixture. In such plants particular type can be maintained only by vegetative method or by careful crossbreeding with the same variety.
3. Double dahlias when grown from seeds show a wide range of colour and also a mixture single, and semi-double. Apple pear, peach, mango etc., do not grow true to type from seeds.
4. Vegetative method of propagation results in earlier flowering and fruiting than those raised from seeds. Vegetative part of a fruiting plant is mature to bear, while a seedling take few years before the shoots ripen to produce flowers and fruits. Seedlings of *Amherstia nobilis* or *Brownea* ariza require 7-8 years to flower but a layer starts flowering in the second year.
5. Plants are also vegetatively propagated to increase their resistance or to develop immunity to particular disease or pest.
6. A well-rooted vegetative part of a plant can adapt more readily to new environment and has greater possibility to flower and fruit than a seedling.

Many parts of plant are capable of giving rise to new plants under natural condition.

Bulbs: Bulbs are underground-modified stem in which the central axis is much shortened and fleshy leaf scales are closely pressed. Amaryllis, Crinum, Hymenocallis, Hemerocallis and Haemanthus, are usually multiplied by bulbs which arise from the main bulb. Cooperanthes, Zephyranthes and tube rose are also manly grown from bulbs.

Corms: Gladiolus produces new corms and cormels on old ones, which are used for multiplication.

Rhizomes, tubers and fleshy roots. Rhizomes are defined as more or less cylindrical branches growing laterally or upward through the soil.

Many ornamental plants such as *Calathea, Anthurium, Alocasia, Alpinia, Hedychium, Heliconia, Gloriosa, Canna* produce rhizomes or rhizomatours stem with buds on it which can be cut into pieces and each one will produce a new plant. Root tubers of dahlia are storage organs and shoots arise from stem attached to the tubers.

Runner: When a slender stem grows out of a crown and trails along the ground it is called a runner or stolon. Chlorophytum, Episcia send out stolons and produce young plants from the nodes. It is then detached and grown separately.

Offsets: Many ornamental plants are propagated by this method. Sanseviera and *Agave americana* send out branches terminating in rosette of leaves which ultimately grow as new plants. These can be separated from the mother plants. Chrysanthemums also produce offsets, which are detached for multiplication of plants. Gerbera also sends out suckers on offsets.

Root Suckers: There are some plants, which produce suckers from roots and are detached from the mother plants for multiplication. *Millingtonia hortensis, Clerodendron splendens, Quisqualis indica* are few examples.

Trees: Most of the species of trees; grown in tropical conditions are raised from seeds. The seedlings with taproot grow tall and large and give necessary support against storm. Some species, however, often fail or produce seeds irregularly.

Seed: Species of *Cassia and Poinciana regia, Thespesia populnea, Peltophorum ferrugineum, couroupita guinensis, Acacia auriculi formis, Pithecolobium saman, Saraca indica, Spathodia campanulata,*

Lagerstroemia flos-regineae and many other flowering trees produce seeds freely which germinate in seed compost consisting of garden soil and leafmould, one part each in bed or in seed pans. Seeds or mahogany have spongy seed coat and rot in excess moisture. Seeds of *Polyalthia* do not germinated if they become too dry. Seedlings of *Eucalyptus* and *Jacaranda* grow well in less humidity and temperature and should be sown in the spring.

Cutting: Plumerias root easily from stem cuttings. Cuttings are made in spring as excessive moisture in soil causes rotting. These are planted in sandy soil and light watering is done after callus formation. Plumerias produce large fruits containing many seeds but the seedlings do not flower before four years. Cutting from mature terminal shoot start flowering within few months. *Gliricidia* are also propagated from cuttings. In the case of *Spathodia campanulata* in addition to seed propagation, root suckers are also used.

Air layer and ground layer. *Gustavia augusta, Brownea ariza and B. grandiceps* though produce few fruits every year but the seedlings are very slow growing and are also propagated by air or ground layering. *Ficus elastica and Ficus krishnae* and *F.benjamina* are easily propagated from air layer.

Shrubs

Seeds: Few species of shrubs which produce seeds and the seedlings flower in one or two years without changing the character, are raised from seeds *e.g., Galphimia gracilis, Cassia glauca, Cassia didymobotrya, Solanum macranthum, Tecoma stans, Sophora tomentosa, Brya ebenus, Callindra speciosa Bauhinia acuminata, B. tomentosa and Caesalpinia pulcherrima speciosa, Ochna wightiana, O. squarrosa, Carissa carundas, Jacquinia ruscifolia, Bauhinia galpinii, Portlandia grandiflora* and *Calliandra speciosa* do not root easily from cutting and layering and so seed multiplication is commonly practised.

All species and varieties of *Acalypha, Angelonia, Aralia, Asystasia, Buddleia, Cestrum, Daedalacanthus, Eranthemum, Graprophyllum, Justicia, Lagerstroemia indica, Malvaviscus, Pentas, Poinsettia, Russelia, Jasminum sambc, Brugmansia, Stachytarpheta and Thunbergia are very easily propagated from cuttings.* In *Aralia, Eranthemum, Pentas, Aphelandra* and *Daedalacanthus*, tip cuttings produce quicker and better roots.

Green House Plants: Greenhouse plants thrive in shade and grow well in high humidity especially during the hot months. These are mostly grown for beautiful foliage though flowering plants like *Achimines* and *Pelargonium* need suitable protection during dry and hot season.

Seeds: Plants can be raised from seeds of *Coleus* but to maintain a particular variety vegetative propagation has to be done.

Stem Cutting: All the different species and varieties of *Aglaonema, Philodendron, Anthurium, Scindapsus, Monstera* and *Dieffenbachia,* have succulent stem which are cut into small pieces keeping one node in each and planted in ground or pot. The cuttings are planted horizontally about 1-2cm below the soil with the bud pointing upward. Several species of *Dracaena e.g., Dracaena ugandense, D. victoria, D. deremensis* "Bausii" D. *reflexa* "Variegata" are grown from top cuttings. The erect stem can also be cut into pieces with few nodes in each cutting and planted in soil. Top cuttings produce well-shaped plants after root formation, while shoots from lower cuttings have slower growth. *Episcia, Syngonium, Gynura, Fittonia, Pellionia, Coleus, Zebrina Pendula, Setcrasea Purpurea, Perperomia, Pillea and Strobilanthes dyerianus* are propagated from stem cuttings. It is safe to plant these soft cuttings in sand only. After root formation it can be transferred in pot or ground. Foliage begonias and African violets can be propagated from leaf cuttings but division is a safe method under our climatic conditions.

Species of ferns, *e.g., Nephrolepis, Pteris, Ptyrogramma, Polypodium* and *Davallia* are divided with crown and rhizome. All palms produce seeds in suitable climate after attaining maturity. *Pritchardia, Caryota, Livistonia* and *Oreodoxa* often for seeds in Calcutta.

Propagation-Bulbs, Tubers

The bulbous plants are divided into two categories *i.e.*, hardy and tender bulbs. The hardy types can be left in the ground and are separated after 2-3 years for their multiplication (*e.g.* Amaryllis, tuberose, canna, zephyranthes, crinum, etc.). On the contrary, the tender types cannot be left in the ground after flowering and are to be lifted from the ground on maturity. After their treatment against diseases, these are to be stored in cold storage or cool places during off season for next year planting *e.g.* Gladiolus, Narcissus, Daffodils, Freesia, Dahlia, etc.

Propagation: The following are the common methods employed for propagation of bulbous plants;

i. *Off-sets, Cormlets or Bulblets:* Mother bulbs produce many bulblets, which are separated and planted for raising new crop. The intensity of bulblet production is a varietal character. Some varieties are very prolific bulblet producer whereas some produce sparse bulblets. It takes about 1-2 years for growing a bulblet into a full bulb.

ii. *Division:* It is followed in tubers, rhizomes and corms, which are divided carefully (in such a way that each piece contains at least one vegetative bud) and planted in the soil.

iii. *Terminal Cuttings:* These cuttings are mostly employed to propagate Dahlia and Rex begonia.

iv. *Seed:* This method is employed by the breeders to create new varieties *e.g.* Gladiolus, Dahlia, Freesia, Amaryllis, Lilies, etc.

Lifting of Bulbs and their Storage

Tender bulbs like Gladiolus, Dahlia, Narcissus, Daffodils, etc. are to be dug from the soil 10-12 weeks after flowering has been finished. Before lifting the bulbs, water is withheld. After digging, bulds are dried in shade for few days. Then these are treated with 0.2% Bavistin solution for 30 minutes and thereafter stored. Hardy bulbs are separated after 2-3 years and are again planted in the planting season.

Begonia sp. (Begoniaceae)

Begonias are grouped into three main classes according to root types *i.e.* rhizomatous (*Begonia red*), tuberous (*B. tuberhybrida*) and fibrous (*B. semperflorens*). Out of three types rhizomatous and fibrous rooted are grown in the plains whereas tuberous rooted can be grown successfully in hills. The important varieties of rhizomatous begonia are: Peace Majesty Silver Queen, Emperor Mikado, Can Can, Crimson Glory, Black Knight, Curly Star Dust, Dew Drop etc. The important varieties of fibrous rooted are: Goldilock, Bo-peep Little Gem, Ballet, Lady Frances, Pink Jewel, Flamengo, Cindrella, Charm, Silver Star, Pink Camellia etc. The tuberous rooted begonias are valued for their attractive flowers and are grouped into Rose form, Camellia flowered, Carnation flowered, Daffodil flowered or Picotee.

Canna Indica (Cannaceae)

Canna is a rhizomatous plant and is very easily propagated by dividing the rhizomes into 10-15 cm pieces during end of June.

Dahlia: Dahlia Variables (Compositae)

Propagation of dahlia is done by seeds, division of tubers and terminal cuttings. The propagation through seeds is easy and the best way to achieve striking mixtures of colourful flowers which are sown in September-October. The varieties which are being marketed in India are Giant Exhibition mixed, Dwarf Double Red Skin, Coltners hybrids, and Border Jewels, Rigoletto, and Citation. Tubers are stored during summer in cool place or in refrigerator. After careful separation, tubers are planted directly in field in August.

The care with a vegetative bud. The terminal cutting should be solid and made in September-October from plants and after treatment with seradix-I are planted in sand. It takes about 2-3 weeks for rooting and after that they are transplanted inpots (25-30cm) or in beds.

Vegetables

Most seeds normally remain viable for 2 or 3 years if stored under good conditions. To grow best vegetables either for home garden or for growing commercially try only certified seeds with, trueness to type and freedom from certain diseases. Good seeds are clean, viable, free from disease and true to the name variety. Therefore, buy only from seed firm of known integrity. High yielding, high price seeds should have 90% germination. Seeds of 50% germination is very poor.

Germination of Seeds: For germination of seeds adequate moisture, temperature and aeration are essential. The requirement of temperature for various vegetable seeds varies markedly. Some seeds do not germinate at low temperature while some others at high. Usually germination is optimum in between 40^0 F and 60^0 F. Seeds absorb moisture and swell and vital activities starts. Respiration begins and energy is supplied and this requires oxygen. Aeration is essential to supply oxygen. If supply of water is more or over wet aeration is poor and may hinder in germination.

Seed Treatments: Vegetable seeds are usually grown in nursery beds or in boxes. There are three types of seed treatments to control diseases.

1. Disinfestation
2. Disinfection.
3. Seed protection.

The first eliminates organisms present on the surface of seeds. Calcium hypochlorite, mercuric chloride and bromide water may be used. The second disinfectant eliminates organisms present within the seeds. For this hot water, formaldehyde and mercuric chloride are effective. In hot water treatment dry seeds are immersed in hot water 45 to 55^0 C for 10 to 15 minutes.

The third treatment protectants are fungicides to protect seeds from soil fungi. Nursery soil should be drenched before sowing of seeds. Commercial materials are also now available to treat seeds like Captan, Cerason, Thiram, Bavistin, Agrosan GN and antibiotics such as Agrimycin (0.01%), Streptocyline (0.01%) etc. To secure healthy seedling the seed treatments should be invariably done. If the outside nursery bed contain worms, maggot and such insects a combination of insecticides and fungicides should be applied like Aldrin, Dieldrin with Captan, Thiram etc. All chemicals should be handled carefully according to directions of the manufacturer.

If boxes are used for growing seedling the soil should be disinfected by heating the soil to 180 F for atleast one hour. If nursery beds are used the soil may be disinfected by 40% Formalin. One part formaldehyde (40%) mixed in 50 parts of water and saturates the soil. After 24 hours of treatments the seeds may be sown. If captan and formaldehyde is used, the nursery soil may be drenched about 10-12 days before seed sowing.

Preparation of Nursery Beds: The nursery beds should be prepared before the soil treatments. The nursery beds should be of one meter width and length four to five meter. The soil should be cultivated to a find tilth. All weeds, stones etc. should be removed. The beds should be raised to 15 to 20 cm. heights. If necessary, shade or shelter may be provided. Usually total bed area of 80-100 sq.m. is sufficient for most of transplanted seedlings to cover one hectare.

NPK nursery mixture may be applied along with compost on FYM during preparation of soil. After complete preparation and levelling the beds should he given soil treatment as mentioned above.

Sowing of Seeds

The vegetable seeds should be sown in nursery beds in lines at 1.5-2 cm. below surface soiled at a distance of 5-6 cm. The distance from line to line should be 10-15 cm. After sowing cover the seeds with sieved compost very lightly. The beds should be watered with a sprinkler. The above distance from plant to plant or row to row may be changed according to size of seeds, kind of vegetables and type of seedlings. Similarly the depth of covering with compost varies with the kind of seed. Very fine seeds may be dusted over the nursery bed. Covering of other seeds may be one to two times their minimum diameter. During germination and after sprinkle water in the beds but no excessive. Over watering creates high humidity and poor aeration and contributes to "damping off" disease.

To transplant the vegetable seedlings to its permanent location the seedlings should be stocky, healthy and vigorous rather than spindly, weak and elongated plants. Moderate temperatures produce stocky growth. To harden some vegetable seedlings, seedlings may be transplanted one or two times to another nursery bed. In large scale condition watering may be withheld two to three days ahead of transplanting to permanent site.

The young seedlings in the nursery bed should protected against diseases and pests. Regular spray to young plants with Malathion 50 EC or Dithane M-45 or Sevin (0.2%) similar material at 10-15 days interval is necessary. Most of the vegetable seedlings in the nursery bed is ready for transplanting when the seedlings are 4 to 6 weeks old. During uprooting every care should be taken not to damage roots. Light watering should be given at least 12 hours before lifting.

Spece Crops

Black Pepper

Black pepper (*Piper nigrum*) is a flowering vine in the family Piperaceae, cultivated for its fruit, which is usually dried and used as a spice and seasoning. The same fruit is also used to produce white pepper and green pepper. Black pepper is native to South India and is extensively cultivated there and elsewhere in tropical regions. The fruit, known as a peppercorn when dried, is a small drupe five millimetres in diameter, dark red when fully mature, containing a single seed.

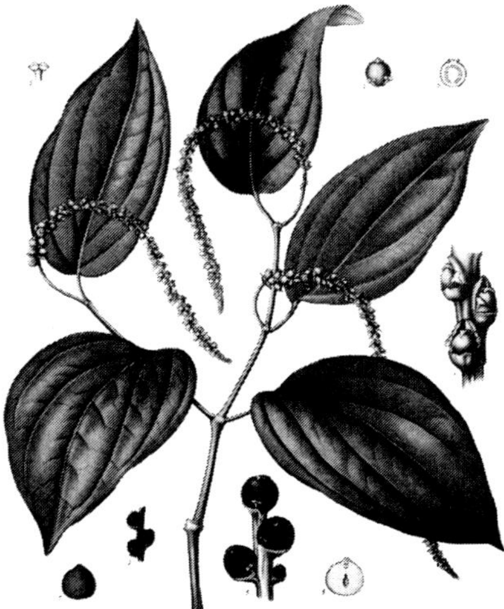

Figure: *Pepper plant with immature peppercorns*

Dried ground pepper is one of the most common spices in European cuisine and its descendants, having been known and prized since antiquity for both its flavour and its use as a medicine. The spiciness of black pepper is due to the chemical piperine. Ground black peppercorn, usually referred to simply as "pepper", may be found on nearly every dinner table in some parts of the world, often alongside table salt.

The word "pepper" is derived from the Sanskrit *pippali*, via the Latin *piper* and Old English *pipor*. The Latin word is also the source of German *pfeffer*, French *poivre*, Dutch *peper*, and other similar forms. In the 16th century, *pepper* started referring to the unrelated New World chile peppers as well. "Pepper" was used in a figurative sense to mean "spirit" or "energy" at least as far back as the 1840s; in the early 20th century, this was shortened to *pep*.

Varieties of Pepper

Black pepper is produced from the still-green unripe berries of the pepper plant. The berries are cooked briefly in hot water, both to clean them and to prepare them for drying. The heat ruptures cell walls in the fruit, speeding the work of browning enzymes during drying. The berries are dried in the sun or by machine for several days, during which the fruit around the seed shrinks and darkens into a thin, wrinkled black layer. Once dried, the fruits are called black peppercorns.

White pepper consists of the seed only, with the fruit removed. This is usually accomplished by allowing fully ripe berries to soak in water for about a week, during which the flesh of the fruit softens and decomposes. Rubbing then removes what remains of the fruit, and the naked seed is dried. Alternative processes are used for removing the outer fruit from the seed, including removal of the outer layer from black pepper produced from unripe berries.

In the U.S., white pepper is often used in dishes like light-coloured sauces or mashed potatoes, where ground black pepper would visibly stand out. There is disagreement regarding which is generally spicier. They do have differing flavours due to the presence of certain compounds in the outer fruit layer of the berry that are not found in the seed. Green pepper, like black, is made from the unripe berries. Dried green peppercorns are treated in a manner that retains the green colour, such as treatment with sulphur dioxide or freeze-drying. Pickled peppercorns, also green, are unripe berries preserved in brine or vinegar. Fresh, unpreserved green pepper berries, largely unknown in the West, are used in some Asian cuisines, particularly Thai cuisine. Their flavour has been described as piquant and fresh, with a bright aroma. They decay quickly if not dried or preserved.

A rarely seen product called pink pepper or red pepper consists of ripe red pepper berries preserved in brine and vinegar. Even more rarely seen, ripe red peppercorns can also be dried using the same colour-preserving techniques used to produce green pepper. Pink pepper from *Piper nigrum* is distinct from the more-common dried "pink peppercorns", which are the fruits of a plant from a different family, the Peruvian pepper tree, *Schinus molle*, and its relative the Brazilian pepper tree, *Schinus terebinthifolius*. In years past there was debate as to the health safety of pink peppercorns, which is mostly no longer an issue. Sichuan peppercorn is another "pepper" that is botanically unrelated to black pepper.

Peppercorns are often categorised under a label describing their region or port of origin. Two well-known types come from India's Malabar Coast: Malabar pepper and Tellicherry pepper. Tellicherry is a higher-grade pepper, made from the largest, ripest 10% of berries from Malabar plants grown on Mount Tellicherry. Sarawak pepper is produced in the Malaysian portion of Borneo, and Lampong pepper on Indonesia's island of Sumatra. White Muntok pepper is another Indonesian product, from Bangka Island.

The Pepper Plant

The pepper plant is a perennial woody vine growing to four metres in height on supporting trees, poles, or trellises. It is a spreading vine, rooting readily where trailing stems touch the ground. The leaves are alternate, entire, five to ten centimetres long and three to six centimetres broad. The flowers are small, produced on pendulous spikes four to eight centimetres long at the leaf nodes, the spikes lengthening to seven to 15 centimetres as the fruit matures.

Black pepper is grown in soil that is neither too dry nor susceptible to flooding, moist, well-drained and rich in organic matter. The plants are propagated by cuttings about 40 to 50 centimetres long, tied up to neighbouring trees or climbing frames at distances of about two metres apart; trees with rough bark are favoured over those with smooth bark, as the pepper plants climb rough bark more readily. Competing plants are cleared away, leaving only sufficient trees to provide shade and permit free ventilation. The roots are covered in leaf mulch and manure, and the shoots are trimmed twice a year. On dry soils the young plants require watering every other day during the dry season for the first three years. The plants bear fruit from the fourth or fifth year, and typically continue to bear fruit for seven years. The cuttings are usually cultivars, selected both for yield and quality of fruit. A single stem will bear 20 to 30 fruiting spikes. The harvest begins as soon as one or two berries at the base of the spikes begin to turn red, and before the fruit is mature, but when full grown and still hard; if allowed to ripen, the berries lose pungency, and ultimately fall off and are lost. The spikes are collected and spread out to dry in the sun, then the peppercorns are stripped off the spikes.

History

Pepper has been used as a spice in India since prehistoric times. J. Innes Miller notes that while pepper was grown in southern Thailand and in Malaysia, its most important source was India, particularly the Malabar Coast, in what is now the state of Kerala. Peppercorns were a much prized trade good, often referred to as "black gold" and used as a form of commodity money. The term "peppercorn rent" still exists today. The ancient history of black pepper is often interlinked with (and confused with) that of long pepper, the dried fruit of closely related *Piper longum*. The Romans knew of both and often referred to either as just "piper". In fact, it was not until the discovery of the New World and of chile peppers that the popularity of long pepper entirely declined. Chile peppers, some of which when dried are similar in shape and taste to long pepper, were easier to grow in a variety of locations more convenient to Europe.

Until well after the Middle Ages, virtually all of the black pepper found in Europe, the Middle East, and North Africa travelled there from India's Malabar region. By the 16th century, pepper was also being grown in Java, Sunda, Sumatra, Madagascar, Malaysia, and elsewhere in Southeast Asia, but these areas traded mainly with China, or used the pepper locally. Ports in the Malabar area also served as a stop-off point for much of the trade in other spices from farther east in the Indian Ocean. Black pepper, along with other spices from India and lands farther east, changed the course of world history. It was in some part the preciousness of these spices that led to the European efforts to find a sea route to India and consequently to the European colonial occupation of that country, as well as the European discovery and colonization of the Americas.

Ancient Times

Black peppercorns were found lodged in the nostrils of Ramesses II, placed there as part of the mummification rituals shortly after his death in 1213 BCE. Little else is known about the use of pepper in ancient Egypt, nor how it reached the Nile from India.

Pepper (both long and black) was known in Greece at least as early as the 4th century BCE, though it was probably an uncommon and expensive item that only the very rich could afford. Trade routes of the time were by land, or in ships which hugged the coastlines of the Arabian Sea. Long pepper, growing in the northwestern part of India, was more accessible than the black pepper from further south; this trade advantage, plus long pepper's greater spiciness, probably made black pepper less popular at the time. By the time of the early Roman Empire, especially after Rome's conquest of Egypt in 30 BCE, open-ocean crossing of the Arabian Sea directly to southern India's Malabar Coast was near routine. Details of this trading across the Indian Ocean have been passed down in the *Periplus of the Erythraean Sea*. According to the Roman geographer Strabo, the early Empire sent a fleet of around 120 ships on an annual one-year trip to India and back.

The fleet timed its travel across the Arabian Sea to take advantage of the predictable monsoon winds. Returning from India, the ships travelled up the Red Sea, from where the cargo was carried overland or via the Nile Canal to the Nile River, barged to Alexandria, and shipped from there to Italy and Rome. The rough geographical outlines of this same trade route would dominate the pepper trade into Europe for a millennium and a half to come.

With ships sailing directly to the Malabar coast, black pepper was now travelling a shorter trade route than long pepper, and the prices reflected it. Pliny the Elder's *Natural History* tells us the prices in Rome around 77 CE: "Long pepper is fifteen denarii per pound, while that of white pepper is seven, and of black, four." Pliny also complains "there is no year in which India does not drain the Roman Empire of fifty million sesterces," and further moralises on pepper:

It is quite surprising that the use of pepper has come so much into fashion, seeing that in other substances which we use, it is sometimes their sweetness, and sometimes their appearance that has attracted our notice; whereas, pepper has nothing in it that can plead as a recommendation to either fruit or berry, its only desirable quality being a certain pungency; and yet it is for this that we import it all the way from India! Who was the first to make trial of it as an article of food? and who, I wonder, was the man that was not content to prepare himself by hunger only for the satisfying of a greedy appetite? (*N.H.* 12.14)

Black pepper was a well-known and widespread, if expensive, seasoning in the Roman Empire. Apicius' De re coquinaria, a 3rd-century cookbook probably based at least partly on one from the 1st century CE, includes pepper in a majority of its recipes. Edward Gibbon wrote, in *The History of the Decline and Fall of the Roman Empire*, that pepper was "a favourite ingredient of the most expensive Roman cookery".

Postclassical Europe

Pepper was so valuable that it was often used as collateral or even currency. The taste for pepper (or the appreciation of its monetary value) was passed on to those who would see Rome fall. It is said that Alaric the Visigoth and Attila the Hun each demanded from Rome a ransom of more than a ton of pepper when they besieged the city in 5th century. After the fall of Rome, others took over the middle legs of the spice trade, first the Persians and then the Arabs; Innes Miller cites the account of Cosmas Indicopleustes, who travelled east to India, as proof that "pepper was still being exported from India in the sixth century". By the end of the Dark Ages, the central portions of the spice trade were firmly under Islamic control. Once into the Mediterranean, the trade was largely monopolised by Italian powers, especially Venice and Genoa. The rise of these city-states was funded in large part by the spice trade. A riddle authored by Saint Aldhelm, a 7th-century Bishop of Sherborne, sheds some light on black pepper's role in England at that time:

I am black on the outside, clad in a wrinkled cover,

Yet within I bear a burning marrow.

I season delicacies, the banquets of kings, and the luxuries of the table,

Both the sauces and the tenderized meats of the kitchen.

But you will find in me no quality of any worth,

Unless your bowels have been rattled by my gleaming marrow.

It is commonly believed that during the Middle Ages, pepper was used to conceal the taste of partially rotten meat. There is no evidence to support this claim, and historians view it as highly unlikely: in the Middle Ages, pepper was a luxury item, affordable only to the wealthy, who certainly had unspoiled meat available as well. Similarly, the belief that pepper was widely used as a preservative is questionable: it is true that piperine, the compound that gives pepper its spiciness, has some antimicrobial properties, but at the concentrations present when pepper is used as a spice, the effect is small. Salt is a much more effective preservative, and salt-cured meats were common fare, especially in winter. However, pepper and other spices probably did play a role in improving the taste of long-preserved meats.

Its exorbitant price during the Middle Ages—and the monopoly on the trade held by Italy—was one of the inducements which led the Portuguese to seek a sea route to India. In 1498, Vasco da Gama became the first European to reach India by sea; asked by Arabs in Calicut (who spoke Spanish and Italian) why they had come, his representative replied, "we seek Christians and spices." Though this first trip to India by way of the southern tip of Africa was only a modest success, the Portuguese quickly returned in greater numbers and used their superior naval firepower to eventually gain complete control of trade on the Arabian sea. This was the start of the first European empire in Asia, given additional legitimacy (at least from a European perspective) by the 1494 Treaty of Tordesillas, which granted Portugal exclusive rights to the half of the world where black pepper originated.

The Portuguese proved unable to maintain their stranglehold on the spice trade for long. The old Arab and Venetian trade networks successfully smuggled enormous quantities of spices through the patchy Portuguese blockade, and pepper once again flowed through Alexandria and Italy, as well as around Africa. In the 17th century, the Portuguese lost almost all of their valuable Indian Ocean possessions to the Dutch and the English. The pepper ports of Malabar fell to the Dutch in the period 1661–1663. As pepper supplies into Europe increased, the price of pepper declined (though the total value of the import trade generally did not). Pepper, which in the early Middle Ages had been an item exclusively for the rich, started to become more of an everyday seasoning among those of more average means. Today, pepper accounts for one-fifth of the world's spice trade.

China

It is possible that black pepper was known in China in the 2nd century BCE, if poetic reports regarding an explorer named Tang Meng are correct. Sent by Emperor Wu to what is now southwest China, Tang Meng is said to have come across something called *jujiang* or "sauce-betel". He was told it came from the markets of Shu, an area in what is now the Sichuan province. The traditional view among historians is that "sauce-betel" is a sauce made from betel leaves, but arguments have been made that it actually refers to pepper, either long or black. In the 3rd century CE, black pepper made its first definite appearance in Chinese texts, as *hujiao* or "foreign pepper". It does not appear to have been widely known at the time, failing to appear in a 4th-century work describing a wide variety of spices from beyond China's southern border, including long pepper.

By the 12th century, however, black pepper had become a popular ingredient in the cuisine of the wealthy and powerful, sometimes taking the place of China's native Sichuan pepper (the tongue-numbing dried fruit of an unrelated plant).

Marco Polo testifies to pepper's popularity in 13th-century China when he relates what he is told of its consumption in the city of Kinsay (Zhejiang): "... Messer Marco heard it stated by one of the Great Kaan's officers of customs that the quantity of pepper introduced daily for consumption into the city of Kinsay amounted to 43 loads, each load being equal to 223 lbs." Marco Polo is not considered a very reliable source regarding China, and this second-hand data may be even more suspect, but if this estimated 10,000 pounds (4,500 kg) a day for one city is anywhere near the truth, China's pepper imports may have dwarfed Europe's.

Pepper as a Medicine

Like all eastern spices, pepper was historically both a seasoning and a medicine. Long pepper, being stronger, was often the preferred medication, but both were used. Black peppercorns figure in remedies in Ayurveda, Siddha and Unani medicine in India. The 5th century *Syriac Book of Medicines* prescribes pepper (or perhaps long pepper) for such illnesses as constipation, diarrhea, earache, gangrene, heart disease, hernia, hoarseness, indigestion, insect bites, insomnia, joint pain, liver problems, lung disease, oral abscesses, sunburn, tooth decay, and toothaches.

Various sources from the 5th century onward also recommend pepper to treat eye problems, often by applying salves or poultices made with pepper directly to the eye. There is no current medical evidence that any of these treatments has any benefit; pepper applied directly to the eye would be quite uncomfortable and possibly damaging.

Pepper has long been believed to cause sneezing; this is still believed true today. Some sources say that piperine irritates the nostrils, causing the sneezing; some say that it is just the effect of the fine dust in ground pepper, and some say that pepper is not in fact a very effective sneeze-producer at all. Few if any controlled studies have been carried out to answer the question.

Pepper is eliminated from the diet of patients having abdominal surgery and ulcers because of its irritating effect upon the intestines, being replaced by what is referred to as a bland diet. Pepper is sometimes used to stop light bleeding in restaurant kitchens.

Flavour

Pepper gets its spicy heat mostly from the piperine compound, which is found both in the outer fruit and in the seed. Refined piperine, milligram-for-milligram, is about one per cent as hot as the capsaicin in chile peppers. The outer fruit layer, left on black pepper, also contains important odour-contributing terpenes including pinene, sabinene, limonene, caryophyllene, and linalool, which give citrusy, woody, and floral notes. These scents are mostly missing in white pepper, which is stripped of the fruit layer. White pepper can gain some different odours (including musty notes) from its longer fermentation stage.

Pepper loses flavour and aroma through evaporation, so airtight storage helps preserve pepper's original spiciness longer. Pepper can also lose flavour when exposed to light, which can transform piperine into nearly tasteless isochavicine. Once ground, pepper's aromatics can evaporate quickly; most culinary sources recommend grinding whole peppercorns immediately before use for this reason. Handheld pepper mills (or "pepper grinders"), which mechanically grind or crush whole peppercorns, are used for this, sometimes instead of pepper shakers, dispensers of pre-ground pepper. Spice mills such as pepper mills were found in European kitchens as early as the 14th century, but the mortar and pestle used earlier for crushing pepper remained a popular method for centuries after as well.

World Trade

Peppercorns are, by monetary value, the most widely traded spice in the world, accounting for 20 percent of all spice imports in 2002. The price of pepper can be volatile, and this figure fluctuates a great deal year to year; for example, pepper made up 39 percent of all spice imports in 1998. By weight, slightly more chile peppers are traded worldwide than peppercorns. The International Pepper Exchange is located in Kochi, India.

Vietnam has recently become the world's largest producer and exporter of pepper (85,000 long tons in 2003). Other major producers include Indonesia (67,000 tons), India (65,000 tons), Brazil (35,000 tons), Malaysia (22,000 tons), Sri Lanka (12,750 tons), Thailand, and China. Vietnam dominates the export market, using almost none of its production domestically. In 2003, Vietnam exported 82,000 tons of pepper, Indonesia 57,000 tons, Brazil 37,940 tons, Malaysia 18,500 tons, and India 17,200 tons.

Cardamom

Figure: *True Cardamom (Elettaria cardamomum)*

The name cardamom (sometimes written cardamon) is used for heba within two genera of the ginger family Zingiberaceae, namely *Elettaria* and *Amomum*.

Types of Cardamom and their Distribution

The two main *genera* of the ginger family that are named as forms of cardamom are distributed as follows:

- *Elettaria* (commonly called cardamom, green cardamom, or true cardamom) is distributed from India to Malaysia.
- *Amomum* (commonly known as black cardamom, brown cardamom, Kravan, Java cardamom, Bengal cardamom, Siamese cardamom, white or red cardamom) is distributed mainly in Asia and Australia.

Uses

All the different cardamom species and varieties are used mainly as cooking spices and as medicines. In general,

- *Elettaria cardamomum* (the usual type of cardamom) is used as a spice, a masticatory, and in medicine; it is also sometimes smoked; it is used as a food plant by the larva of the moth *Endoclita hosei.*
- *Amomum* is used as an ingredient in traditional systems of medicine in China, India, Korea, Japan, and Vietnam.
- Can be used as a traditional flavouring to Turkish coffee.
- Is often used in the traditional Indian tea, or chai, especially with milk and sugar.

Uses in Cuisines Around the World

Cardamom has a strong, unique taste, with an intensely aromatic fragrance. It is a common ingredient in Indian cooking, and is often used in baking in Scandinavia.

One of the most expensive spices by weight, little is needed to impart the flavour. Cardamom is best stored in pod form, because once the seeds are exposed or ground, they quickly lose their flavour.

However, high-quality ground cardamom is often more readily (and cheaply) available, and is an acceptable substitute. For recipes requiring whole cardamom pods, a generally accepted equivalent is 10 pods equals 1½ teaspoons of ground cardamom.

In Traditional Medicine

In India, green cardamom (*A. subulatum*), or "elaichi," is broadly used to treat infections in teeth and gums, to prevent and treat throat troubles, congestion of the lungs and pulmonary tuberculosis, inflammation of eyelids and also digestive disorders. It is also reportedly used as an antidote for both snake and scorpion venom.

Species in the genus *Amomum* are also used in traditional Indian medicine. Among other species, varieties and cultivars, *Amomum villosum* is used in traditional Chinese medicine to treat stomach-aches, constipation, dysentery, and other digestion problems. "Tsaoko" cardamom is cultivated in Yunnan, China, both for medicinal purposes and as a spice.

Ginger

Ginger is commonly used as a spice in cuisines throughout the world. Though commonly referred to as a root, it is actually the rhizome of the monocotyledonous perennial plant *Zingiber officinale.* Originating in southern China, cultivation of ginger spread to India, Southeast Asia, West Africa, and the Caribbean.

Chemistry

Ginger contains up to 3% of an essential oil that causes the fragrance of the spice. The main constituents are sesquiterpenoids with (-)-zingiberene as the main component. Lesser amounts of other sesquiterpenoids (a-sesquiphell-andrene, bisabolene and farnesene) and a small monoterpenoid fraction (a-phelladrene, cineol, and citral) have also been identified.

The pungent taste of ginger is due to nonvolatile phenylpropanoids (particularly gingerol and zingerone) and diarylheptanoids (gingeroles and shoagoles); the latter are more pungent and form from the former when ginger is dried. With a specific procedure is used for cooking, where ginger root acquires a soda form and transforms gingerol into zingerone, which is less pungent and has a spicy-sweet aroma.

Culinary Uses

Young ginger roots are juicy and fleshy with a very mild taste. They are often pickled in vinegar or sherry as a snack or just cooked as an ingredient in many dishes. They can also be stewed in boiling water to make ginger tea, to which honey is often added as a sweetener. Mature ginger roots are fibrous and nearly dry. The juice from old ginger roots is extremely potent and is often used as a spice in Chinese cuisine to flavour dishes such as in seafood and mutton.

Ginger is also candied, is used as a flavoring for candy, cookies, crackers and cake, and is the main flavour in ginger ale, a sweet, carbonated, non-alcoholic beverage, as well as the similar, but somewhat spicier beverage ginger beer. A ginger-flavoured liqueur called Canton is produced in the Guangdong province of China; it is advertised to be based on a recipe created for the rulers of the Qing Dynasty and made from six different varieties of ginger. Green ginger wine is a ginger flavoured wine produced in the United Kingdom by Crabbie's and Stone's and traditionally sold in a green glass bottle. Ginger is also used as a spice added to hot coffee and tea.

In Japan, ginger is pickled to make beni shoga and gari or grated and used raw on tofu or noodles.

In Western cuisine, ginger is traditionally restricted to sweet foods, such as ginger ale, gingerbread, ginger snaps, ginger cake and ginger biscuits.

Powdered dry ginger root (ground ginger) is typically used to add spiciness to gingerbread and other recipes. Ground and fresh ginger taste quite different and ground ginger is a particularly poor substitute for fresh ginger. Fresh ginger can be successfully substituted for ground ginger and should be done at a ratio of 6 parts fresh for 1 part ground. You generally achieve better results by substituting only half the ground ginger for fresh ginger. In Myanmar, ginger is used in a salad dish called *gyin-tho*, which consists of shredded ginger preserved in oil, and a variety of nuts and seeds.

In traditional Korean Kimchi, ginger is minced finely and added into the ingredients of the spicy paste just before the fermenting process. In India, ginger is used in all sub-varieties of the Indian cuisines. In south India, ginger is used in the production of a candy called Inji-murappa ("ginger candy" from Tamil). This candy is mostly sold by vendors to bus passengers in bus stops and in small tea shops as a locally produced item. Candied ginger is also very famous around these parts.

Additionally, in Tamil Nadu, especially in the Tanjore belt, a variety of ginger which is less spicy is used when tender to make fresh pickle with the combination of lemon juice or vinegar, salt and tender green chillies. This kind of pickle was generally made before the invention of refrigeration and stored for a maximum of 4-5 days. The pickle gains a mature flavour when the juices cook the ginger over the first 24 hours. Ginger is used in the curries of North Indian food or cooked into the food. In South East Asia, the flower of a type of ginger is used in cooking. This unopened flower is known in the Malay language as Bunga Kantan, and is used in salads and also as garnish for sour-savoury soups, like Assam Laksa.

Ginger has a sialagogue action, stimulating the production of saliva.

Medicinal Uses

One medical research study had results indicating that ginger might be an effective treatment for nausea caused by motion sickness or other illness, The study however, failed to show a significant difference between ginger and a placebo. There are several proposed mechanisms of action for the anti-emetic properties of ginger but there is not yet conclusive support for any particular model.

Modern research on nausea and motion sickness used approximately 1 gram of ginger powder daily. Though there are claims for efficacy in all causes of nausea, the PDR recommends against taking ginger root for morning sickness commonly associated with pregnancy due to possible mutagenic effects. Nevertheless, Chinese women traditionally have taken ginger root during pregnancy to combat morning sickness. The Natural Medicines Comprehensive Database (compiled by health professionals and pharmacists), states that ginger is likely safe for use in pregnancy when used orally in amounts found in foods. Ginger ale and ginger beer have been recommended as "stomach settlers" for generations in countries where the beverages are made. Ginger water was commonly used to avoid heat cramps in the United States in the past.

In Western-hemisphere nations, powdered dried ginger root is made into capsules and sold in pharmacies for medicinal use. In the US, ginger is not approved by the FDA for the treatment or cure of any disease. Ginger is instead sold as an unregulated dietary supplement. In India, ginger is applied as a paste to the temples to relieve headache. In Myanmar, ginger and local sweet (Htan nyat) which is made from palm tree juice are boiled together and taken to prevent the Flu. A hot ginger drink (made with sliced ginger cooked in sweetened water or a Coca-Cola-like drink) has been reported as a folk medicine for common cold. Ginger has also historically been used in folk medicine to treat inflammation, although medical studies as to the efficacy of ginger in decreasing inflammation have shown mixed results.

There are several studies that demonstrate a decrease in joint pain from arthritis after taking ginger, though the results have not been consistent from study to study. It may also have blood thinning and cholesterol lowering properties, making it theoretically effective in treating heart disease; while early studies have shown some efficacy, it is too early to determine whether further research will bear this out. The medical form of ginger historically was called "Jamaica ginger"; it was classified as a stimulant and carminative, being much used for dyspepsia and colic. It was also frequently employed to disguise the taste of nauseous medicines. The tea brewed from this root was an old-fashioned remedy for colds. The characteristic odour and flavour of ginger root is caused by a mixture of zingerone, shoagoles and gingerols, volatile oils that compose about 1%–3% by weight of fresh ginger. The gingerols have analgesic, sedative, antipyretic, antibacterial, and GI tract motility effects.

Ginger is on the GRAS list from FDA. However, like other herbs, ginger may be harmful because it may interact with other medications, such as warfarin; hence, a physician or pharmacist should be consulted before taking the herb. Ginger is also contraindicated in people suffering from gallstones, because the herb promotes the release of bile from the gallbladder.

Ginger Allergies

Some people are allergic to ginger. Generally, this is reported as having a gaseous component. This may take the form of flatulence, or it may take the form of an extreme constriction or tightening in the throat necessitating uncontrollable burping to relieve the pressure.

Horticulture

Ginger produces clusters of white and pink flower buds that bloom into yellow flowers. Because of the aesthetic appeal and the adaptivity of the plant to warm climates, ginger is often used as landscaping around subtropical homes. It is a perennial reed-like plant with annual leafy stems, three to four feet high.

Historical methods of gathering the root describes, when the stalk withers, it is immediately scalded, or washed and scraped, in order to kill it and prevent sprouting. The former method, applied generally to the older and poorer roots, produces Black Ginger; the latter, gives White Ginger. The natural colour of the "white" scraped ginger is a pale buff—it is often whitened by bleaching or liming, but generally at the expense of some of its real value.

References in popular culture:

- To members of the Race, an alien species in Harry Turtledove's best-selling novel series Worldwar, ginger is a highly addictive, psychoactive drug, with an effect similar to that of cocaine or PCP in humans.
- In Cockney rhyming slang, *ginger* is a derogatory euphemism for *homosexual*. The original slang rhymed *queer* with *ginger beer*.
- In the west of Scotland (particularly Glasgow), *ginger* is a term for any carbonated soft drink.
- Before the First World War, it was common for mounted regiments to receive large vats of root ginger before public ceremonies, which were peeled and cut into suppositories for the horses. The burning sensation made the horses hold their tails up; this practice is called Figging or feaguing.
- Ginger is also a common slang term in Great Britain for red-haired individuals, while In North America (USA and Canada) terms involving the word "carrot" are much more common and in Australian English slang, a red-head is a *bluey / blooie*.

Similar Species

Myoga (*Zingiber mioga* Roscoe) appears in Japanese cuisine; the flower buds are the part eaten.

Another plant in the *Zingiberaceae* family, galangal, is used for similar purposes as ginger in Thai cuisine. Galangal is also called Thai ginger. Also referred to as galangal, fingerroot (*Boesenbergia rotunda*), or Chinese ginger or the Thai *krachai*, is used in cooking and medicine.

A dicotyledonous native species of eastern North America, *Asarum canadense*, is also known as "wild ginger", and its root has similar aromatic properties, but it is not related to true ginger and should not be used as a substitute because it contains the carcinogen aristolochic acid. This plant is also a powerful diuretic, or urinary stimulator. It is part of the Aristolochiaceae family.

Turmeric

Turmeric (*Curcuma longa*) is a member of the ginger family, Zingiberaceae. It's also called tumeric or kunyit in some Asian countries.

Figure: *Curcuma longa*

Its dried roots are ground into a deep yellow spice commonly used in curries and other South Asian cuisine. Its active ingredient is curcumin and it has an earthy, bitter, peppery flavour. Sangli, a town in the southern part of the Indian state of Maharashtra, is the largest and most important trading centre for turmeric in Asia or perhaps in the entire world.

Uses

Food: Turmeric has found application in canned beverages, baked products, dairy products, ice cream, yogurt, yellow cakes, biscuits, popcorn-colour, sweets, cake icings, cereals, sauces, gelatins, etc. It is a significant ingredient in most commercial curry powders.

Turmeric (coded as E100 when used as a food additive) is used to protect food products from sunlight. The oleoresin is used for oil-containing products. The curcumin/polysorbate solution or curcumin powder dissolved in alcohol is used for water containing products. Over-colouring, such as in pickles, relishes and mustard, is sometimes used to compensate for fading.

In combination with annatto (E160b), turmeric has been used to colour cheeses, dry mixes, salad dressings, winter butter and margarine. Turmeric is also used to give a yellow colour to some prepared mustards, canned chicken broths and other foods (often as a much cheaper replacement for saffron). Momos (Nepali meat dumplings), a traditional dish in South Asia, are spiced with turmeric.

Medicine: In the Ayurvedic medicine, turmeric is thought to have many medicinal properties and many in India use it as a readily available antiseptic for cuts and burns. Whenever there is a cut or a bruise, the home remedy is to reach for turmeric powder. Ayurvedic doctors say it has fluoride which is essential for teeth. It is also used as an antibacterial agent.

It is taken in some Asian countries as a dietary supplement, which allegedly helps with stomach problems and other ailments. It is popular as a tea in Okinawa, Japan. It is currently being investigated for possible benefits in Alzheimer's disease, cancer and liver disorders.

It is only in recent years that Western scientists have increasingly recognised the medicinal properties of turmeric. According to a 2005 article in the Wall Street Journal titled, "Common Indian Spice Stirs Hope," research activity into curcumin, the active ingredient in turmeric, is exploding. Two hundred and fifty-six curcumin papers were published in the past year according to a search of the U.S. National Library of Medicine. Supplement sales have increased 35% from 2004, and the U.S. National Institutes of Health has four clinical trials underway to study curcumin treatment for pancreatic cancer, multiple myeloma, Alzheimer's, and colorectal cancer.

A 2004 UCLA-Veterans Affairs study involving genetically altered mice suggests that curcumin, the active ingredient in turmeric, might inhibit the accumulation of destructive beta amyloids in the brains of Alzheimer's disease patients and also break up existing plaques. "Curcumin has been used for thousands of years as a safe anti-inflammatory in a variety of ailments as part of Indian traditional medicine," Gregory Cole, Professor of medicine and neurology at the David Geffen School of Medicine at UCLA said.

Another 2004 study conducted at Yale University involved oral administration of curcumin to mice homozygous for the most common allele implicated in cystic fibrosis. Treatment with curcumin restored physiologically-relevant levels of protein function. Anti-tumoral effects against melanoma cells have been demonstrated.

Curry Pharmaceuticals, based in North Carolina, is studying the use of a curcumin cream for psoriasis treatment. Another company is already selling a cream based on curcumin called "Psoria-Gold," which shows anecdotal promise of treating the disease. A recent study involving mice has shown that turmeric slows the spread of breast cancer into lungs and other body parts. Turmeric also enhances the effect of taxol in reducing metastasis of breast cancer.

Curcumin is thought to be a powerful antinociceptive (pain-relieving) agent. In the November 2006 issue of *Arthritis & Rheumatism*, a study was published that showed the effectiveness of turmeric in the reduction of joint inflammation, and recommended clinical trials as a possible treatment for the alleviation of arthritis symptoms. It is thought to work as a natural inhibitor of the cox-2 enzyme, and has been shown effective in animal models for neuropathic pain secondary to diabetes, among others.

Cosmetics: Turmeric is currently used in the formulation of some sunscreens. Turmeric paste is used by some Indian women to keep them free of superfluous hair. Turmeric paste is applied to bride and groom before marriage in some places of India, where it is believed turmeric gives glow to skin and keeps some harmful bacteria away from the body. The Government of Thailand is funding a project to extract and isolate tetrahydrocurcuminoids (THC) from turmeric. THCs are colorless compounds that might have antioxidant and skin lightening properties and might be used to treat skin inflammations, making these compounds useful in cosmetics formulations.

Dye: Turmeric makes a poor fabric dye as it is not very lightfast (the degree to which a dye resists fading due to light exposure).

Chemistry: The active substance of turmeric is the polyphenol **curcumin**, also known as C.I. 75300, or Natural Yellow 3. Systematic chemical name is (1*E*,6*E*)-1,7-bis(4-hydroxy-3-methoxyphenyl)-1,6-heptadiene-3,5-dione. It can exist at least in two tautomeric forms, keto and enol. The keto form is preferred in solid phase and the enol form in solution.

Nutmeg: Nutmeg is the actual seed of the tree, roughly egg-shaped and about 20–30 mm long and 15–18 mm wide, and weighing between 5 and 10 grams dried, while mace is the dried "lacy" reddish covering or arillus of the seed. Several other commercial products are also produced from the trees, including essential oils, extracted oleoresins, and nutmeg butter. The pericarp (fruit/pod) is used in Grenada to make a jam called Morne Delice.

In Indonesia, the fruit is sliced finely, cooked and crystallised to make a fragrant candy called *manisan pala* ("nutmeg sweets").

The most important species commercially is the Common or Fragrant Nutmeg *Myristica fragrans*, native to the Banda Islands of Indonesia; it is also grown in the Caribbean, especially in Grenada. Other species include Papuan Nutmeg *M. argentea* from New Guinea, and Bombay Nutmeg *M. malabarica* from India; both are used as adulterants of *M. fragrans* products.

Culinary Uses: Nutmeg and mace have similar taste qualities, nutmeg having a slightly sweeter and mace a more delicate flavour. Mace is often preferred in light-coloured dishes for the bright orange, saffron-like colour it imparts. Nutmeg is nice in cheese sauces and is best grated fresh. In Indian cuisine, nutmeg is used almost exclusively in sweets. It is known as *jaiphal* in most parts of India. It is also used in small quantities in garam masala. In other European cuisine, nutmeg and mace are used especially in potato dishes and in processed meat products; they are also used in soups, sauces and baked goods.

Japanese varieties of curry powder include nutmeg as an ingredient. Nutmeg is a traditional ingredient in mulled cider, mulled wine, and eggnog.

Essential Oils: The essential oil is obtained by the steam distillation of ground nutmeg and is used heavily in the perfumery and pharmaceutical industries. The oil is colourless or light yellow and smells and tastes of nutmeg. It contains numerous components of interest to the oleochemical industry, and is used as a natural food flavouring in baked goods, syrups (*e.g.* Coca Cola), beverages, sweets etc. It replaces ground nutmeg as it leaves no particles in the food.

The essential oil is also used in the cosmetic and pharmaceutical industries for instance in tooth paste and as major ingredient in some cough syrups. In traditional medicine nutmeg and nutmeg oil were used for illnesses related to the nervous and digestive systems. Myristicin and elemicin are believed to be the chemical constituents responsible for the subtle hallucinogenic properties of nutmeg oil. Other known chemical ingredients of the oil are a-pinene, sabinene, a-terpinene and safrole.

Externally, the oil is used for rheumatic pain and, like clove oil, can be applied as an emergency treatment to dull toothache. Put 1–2 drops on a cotton swab, and apply to the gums around an aching tooth until dental treatment can be obtained. In France, it is given in drop doses in honey for digestive upsets and used for bad breath. Use 3–5 drops on a sugar lump or in a teaspoon of honey for nausea, gastroenteritis, chronic diarrhea, and indigestion. Alternatively a massage oil can be created by diluting 10 drops in 10 ml almond oil. This can be used for muscular pains associated with rheumatism or overexertion. It can also be combined with thyme or rosemary essential oils. To prepare for childbirth, massaging the abdomen daily in the three weeks before the baby is due with a mixture of 5 drops nutmeg oil and no more than 5 drops sage oil in 25 ml almond oil has been suggested.

Nutmeg Butter: Nutmeg butter is obtained from the nut by expression. It is semi solid and reddish brown in colour and tastes and smells of nutmeg. Approximately 75% (by weight) of nutmeg butter is trimyristin which can be turned into myristic acid, a 14-carbon fatty acid which can be used as replacement for cocoa butter, can be mixed with other fats like cottonseed oil or palm oil, and has applications as an industrial lubricant.

History. There is some evidence that Roman priests may have burned nutmeg as a form of incense, although this is disputed. It is known to have been used as a prized and costly spice in medieval cuisine. Saint Theodore the Studite (ca. 758– ca. 826), was famous for allowing his monks to sprinkle nutmeg on their pease pudding when required to eat it. In Elizabethan times it was believed that nutmeg could ward off the plague, so nutmeg was very popular. Nutmeg was traded by Arabs during the Middle Ages in the profitable Indian Ocean trade. In the late 15th century, Portugal theoretically took over the Indian Ocean trade, including nutmeg, under the Treaty of Tordesillas with Spain and a separate treaty with the sultan of Ternate. But their control of this trade was always only partial and they remained largely participants, rather than overlords. The authority Ternate held over the nutmeg-growing centre of the Banda Islands was quite limited, and the Portuguese failed to gain a serious foothold in the islands themselves.

The trade in nutmeg later became dominated by the Dutch in the 17th century. The British and Dutch engaged in prolonged struggles and intrigue to gain control of Run island, then the only source of nutmegs. At the end of the Second Anglo-Dutch War the Dutch gained control of Run in exchange for the British controlling New Amsterdam (New York) in North America.

The Dutch managed to establish control over the Banda Islands after an extended military campaign that culminated in the massacre or expulsion of most of the islands' inhabitants in 1621. Thereafter, the Banda Islands were run as a series of plantation estates, with the Dutch mounting annual expeditions in local war-vessels to extirpate nutmeg trees planted elsewhere.

As a result of the Dutch interregnum during the Napoleonic Wars, the English took temporary control of the Banda Islands from the Dutch and transplanted nutmeg trees to their own colonial holdings elsewhere, notably Zanzibar and Grenada. Today, a stylised split-open nutmeg fruit is found on the modern national flag of Grenada. Connecticut gets its nickname ("the Nutmeg State", "Nutmegger") from the legend that some unscrupulous Connecticut traders would whittle "nutmeg" out of wood, creating a "wooden nutmeg" (a term which came to mean any fraud).

World Production: World production of nutmeg is estimated to average between 10,000 and 12,000 tonnes per year with annual world demand estimated at 9,000 tonnes; production of mace is estimated at 1,500 to 2,000 tonnes. Indonesia and Grenada dominate production and exports of both products with a world market share of 75% and 20% respectively. Other producers include India, Malaysia, Papua New Guinea, Sri Lanka and Caribbean islands such as St. Vincent. The principal import markets are the European Community, the United States, Japan and India. Singapore and the Netherlands are major re-exporters.

At one time, nutmeg was one of the most valuable spices. It has been said that in England, several hundred years ago, a few nutmeg nuts could be sold for enough money to enable financial independence for life. The first harvest of nutmeg trees takes place 7–9 years after planting and the trees reach their full potential after 20 years.

Risks and Toxicity: In low doses, nutmeg produces no noticeable physiological or neurological response. Large doses of 30 g or more are dangerous, potentially inducing convulsions, palpitations, nausea, eventual dehydration, and generalized body pain. In amounts of 5–20 g it is a mild to medium hallucinogen, producing visual distortions and a mild euphoria. It is a common misconception that nutmeg contains monoamine oxidase inhibitors (MAOIs).

This is untrue; nutmeg should not be taken in combination with MAOIs but it does not contain them. A test was carried out on the substance which showed that, when ingested in large amounts, nutmeg takes on a similar chemical make-up to MDMA (ecstasy). However, use of nutmeg as a recreational drug is unpopular due to its unpleasant taste and its side effects, including dizziness, flushes, dry mouth, accelerated heartbeat, temporary constipation, difficulty in urination, nausea, and panic. A user will not experience a peak until approximately six hours after ingestion, and effects can linger for up to three days afterwards. A risk in any large-quantity (over 25 g) ingestion of nutmeg is the onset of 'nutmeg poisoning', an acute psychiatric disorder marked by thought disorder, a sense of impending death, and agitation. Some cases have resulted in hospitalization.

Clove

Cloves (*Syzygium aromaticum*, syn. *Eugenia aromaticum* or *Eugenia caryophyllata*) are the aromatic dried flower buds of a tree in the family Myrtaceae. It is native to Indonesia and used as a spice in cuisine all over the world. The name derives from French *clou*, a nail, as the buds vaguely resemble small irregular nails in shape. Cloves are harvested primarily in Zanzibar, Indonesia and Madagascar; it is also grown in India, and Sri Lanka.

The clove tree is an evergreen which grows to a height ranging from 10-20 m, having large oval leaves and crimson flowers in numerous groups of terminal clusters. The flower buds are at first of a pale colour and gradually become green, after which they develop into a bright red, when they are ready for collecting. Cloves are harvested when 1.5-2 cm long, and consist of a long calyx, terminating in four spreading sepals, and four unopened petals which form a small ball in the centre.

Uses

Cloves can be used in cooking either whole or in a ground form, but as they are extremely strong, they are used sparingly. The spice is used throughout Europe and Asia and is smoked in a type of cigarettes locally known as *kretek* in Indonesia and in occasional coffee bars in the West, mixed with marijuana to create marijuana spliffs (joints). Cloves are also an important incense material in Chinese and Japanese culture. Clove essential oil is used in aromatherapy and oil of cloves is widely used to treat toothache in dental emergencies.

Cloves have historically been used in Indian cuisine (both North Indian and South Indian). In the north Indian cuisine, it is used in almost every sauce or side dish made, mostly ground up along with other spices. They are also a key ingredient in tea along with green cardamoms. In the south Indian cuisine, it finds extensive use in the biryani dish (similar to the pilaf, but with the addition of local spice taste), and is normally added whole to enhance the presentation and flavour of the rice. Along with the recreational uses of cloves, they are also said to be a natural anthelmintic.

History

Until modern times, cloves grew only on a few islands in the Maluku Islands (historically called the Spice Islands), including Bacan, Makian, Moti, Ternate, and Tidore. Nevertheless, they found their way west to the Middle East and Europe well before the time of Christ. Archeologists found cloves within a ceramic vessel in Syria along with evidence dating the find to within a few years of 1721 BC.

Cloves, along with nutmeg and pepper, were highly prized in Roman times, and Pliny the Elder once famously complained that "there is no year in which India does not drain the Roman Empire of fifty million sesterces". Cloves were traded by Arabs during the Middle Ages in the profitable Indian Ocean trade. In the late fifteenth century, Portugal took over the Indian Ocean trade, including cloves, due to the Treaty of Tordesillas with Spain and a separate treaty with the sultan of Ternate. The Portuguese brought large quantities of cloves to Europe, mainly from the Maluku Islands. Clove was then one of the most valuable spices, a kg costing around 7 g of gold. The trade later became dominated by the Dutch in the seventeenth century. With great difficulty the French succeeded in introducing the clove tree into Mauritius in the year 1770; subsequently their cultivation was introduced into Guiana, Brazil, most of the West Indies, and Zanzibar, where the majority of cloves are grown today. In Britain in the seventeenth and eighteenth centuries, cloves were worth at least their weight in gold, due to the high price of importing them. The clove has become a commercial 'success', with products including clove drops being released and enjoyed by die-hard clove fans.

Active Compounds

The compound responsible for the cloves' aroma is eugenol. It is the main component in the essential oil extracted from cloves, comprising 72-90%. Eugenol has pronounced antiseptic and anaesthetic properties.

Cinnamon

Cinnamon (*Cinnamomum verum*, synonym *C. zeylanicum*) is a small evergreen tree 10-15 meters (32.8-49.2 feet) tall, belonging to the family Lauraceae, native to Sri Lanka and Southern India. The bark is widely used as a spice. The leaves are ovate-oblong in shape, 7-18 cm (2.75-7.1 inches) long. The flowers, which are arranged in panicles, have a greenish colour, and have a rather disagreeable odour. The fruit is a purple one-centimetre berry containing a single seed.

Its flavour is due to an aromatic essential oil which makes up 0.5 to 1% of its composition. This oil is prepared by roughly pounding the bark, macerating it in sea-water, and then quickly distilling the whole. It is of a golden-yellow colour, with the characteristic odour of cinnamon and a very hot aromatic taste. The pungent taste and scent come from cinnamic aldehyde or cinnamaldehyde and, by the absorption of oxygen as it ages, it darkens in colour and develops resinous compounds. Chemical components of the essential oil include ethyl cinnamate, eugenol, cinnamaldehyde, beta-caryophyllene, linalool and methyl chavicol. The name cinnamon comes from Greek *kinnámômon*, from Phoenician and akin to Hebrew *qinnâmôn*, itself ultimately from a Malaysian language, cf. Malay and Indonesian *kayu manis* "sweet wood".

History

Cinnamon has been known from remote antiquity, and it was so highly prized among ancient nations that it was regarded as a gift fit for monarchs and other great potentates. It was imported to Egypt from China as early as 2000 BC, and is mentioned in the Bible in Exodus 30:23, where Moses is commanded to use both sweet cinnamon and cassia, and in Proverbs 7:17-18, where the lover's bed is perfumed with myrrh, aloe and cinnamon. It is also alluded to by Herodotus and other classical writers. It was commonly used on funeral pyres in Rome, and the Emperor Nero is said to have burned a year's supply of cinnamon at the funeral for his wife Poppaea Sabina, in 65 AD.

In the Middle Ages, the source of cinnamon was a mystery to the Western world. Arab traders brought the spice via overland trade routes to Alexandria in Egypt, where it was bought by Venetian traders from Italy who held a monopoly on the spice trade in Europe. The disruption of this trade by the rise of other Mediterranean powers such as the Mameluk Dynasties and the Ottoman Empire was one of many factors that led Europeans to search more widely for other routes to Asia.

Portuguese traders finally discovered Ceylon (Sri Lanka) at the end of the fifteenth century, and restructured the traditional production of cinnamon by the *salagama* caste. The Portuguese established a fort on the island in 1518, and protected their own monopoly for over a hundred years. Dutch traders finally dislodged the Portuguese by allying with the inland Ceylon kingdom of Kandy. They established a trading post in 1638, took control of the factories by 1640, and expelled all remaining Portuguese by 1658. "The shores of the island are full of it", a Dutch captain reported, "and it is the best in all the Orient: when one is downwind of the island, one can still smell cinnamon eight leagues out to sea" (Braudel 1984, p. 215).

The Dutch East India Company continued to overhaul the methods of harvesting in the wild, and eventually began to cultivate its own trees. The British took control of the island from the Dutch in 1796. However, the importance of the monopoly of Ceylon was already declining, as cultivation of the cinnamon tree spread to other areas, the more common cassia bark became more acceptable to consumers, and coffee, tea, sugar and chocolate began to outstrip the popularity of traditional spices.

Cultivation

Cinnamon is harvested by growing the tree for two years and then coppicing it. The next year a dozen or so shoots will form from the roots. These shoots are then stripped of their bark which is left to dry. Only the thin (0.5 mm) inner bark is used; the outer woody portion is removed, leaving metre long cinnamon strips that curl into rolls ("quills") on drying; each dried quill comprises strips from numerous shoots packed together. These quills are then cut to 5-10 cm long pieces for sale.

Cinnamon comes from Sri Lanka, and the tree is also grown commercially at Tellicherry in southern India, Java, Sumatra, the West Indies, Brazil, Vietnam, Madagascar, Zanzibar, and Egypt. Sri Lanka cinnamon is a very thin smooth bark, with a light-yellowish brown colour, a highly fragrant odour.

Cinnamon and Cassia

The name *cinnamon* is correctly used to refer to Ceylon Cinnamon, also known as "true cinnamon" (from the botanical name *C. verum*). However, the related species Cassia (*Cinnamomum aromaticum*) and Cinnamomum burmannii are sometimes sold labelled as cinnamon, sometimes distinguished from true cinnamon as "Indonesian cinnamon" or, at least for Cassia, "Bastard cinnamon". Ceylon cinnamon, using only the thin inner bark, has a finer, less dense and more crumbly texture, and is considered to be less strong than cassia. Cassia is generally a medium to light reddish brown, is hard and woody in texture, and is thicker (2-3 mm thick), as all of the layers of bark are used. Most of the cinnamon sold in supermarkets in the United States is actually cassia. European health agencies have recently warned against consuming high amounts of cassia, due to a toxic component called coumarin. This is contained in much lower dosages in Ceylon cinnamon and in Cinnamomum burmannii. Coumarin is known to cause liver and kidney damage in high concentrations.

The two barks when whole are easily distinguished, and their microscopic characteristics are also quite distinct. Cinnamon sticks (or quills) have many thin layers and can easily be made into powder using a coffee or spice grinder whereas cassia sticks are much harder, made up of one thick layer, capable of damaging a spice or coffee grinder. It is a bit harder to tell powdered cinnamon from powdered cassia. When powdered bark is treated with tincture of iodine (a test for starch), little effect is visible in the case of pure cinnamon of good quality, but when cassia is present a deep-blue tint is produced, the intensity of the coloration depending on the proportion of cassia.

Cinnamon is also sometimes confused with Malabathrum (*Cinnamomum tamala*) and Saigon Cinnamon (*Cinnamomum loureiroi*).

Uses

Cinnamon bark is widely used as a spice. It is principally employed in cookery as a condiment and flavouring material, being largely used in the preparation of some kinds of desserts, chocolate, spicy candies, tea, hot cocoa and liqueurs. In the Middle East, it is often used in savoury dishes of chicken and lamb. In the United States, cinnamon and sugar are often used to flavour cereals, bread-based dishes, and fruits, especially apples; a cinnamon-sugar mixture is even sold separately for such purposes. Cinnamon can also be used in pickling. Cinnamon bark is one of the few spices which can be consumed directly.

In medicine it acts like other volatile oils and once had a reputation as a cure for colds. It has also been used to treat diarrhea and other problems of the digestive system. Cinnamon is high in antioxidant activity (PMID 16190627, PMID 10077878). The essential oil of cinnamon also has antimicrobial properties (PMID 16104824).

This property may allow cinnamon to extend the shelf life of foods. In the media, “cinnamon” has been reported to have remarkable pharmacological effects in the treatment of type II diabetes.

However, the plant material used in the study (PMID 14633804) was actually cassia, as opposed to true cinnamon. Please refer to cassia’s medicinal uses for more information about its health benefits. Cinnamon has traditionally been used to treat toothache and fight bad breath and its regular use is believed to stave off common cold and aid digestion. Cinnamon is used in the system of Thelemic Magick for the invocation of Apollo, according to the correspondences listed in Aleister Crowley's work *Liber 777.* Cinnamon is also used as an insect repellent. It is widely used when a manufactured insecticide is not wanted or cannot be used because of possible health side effects or allergies.

Allspice

Allspice, also called Jamaica pepper, Myrtle pepper, pimento, or newspice, is a spice which is the dried unripe fruit of the *Pimenta dioica* plant. The name "allspice" was coined by the English, who thought it combined the flavour of several spices, such as salt, chili powder, and garlic.

Flavour

Allspice has a complex aroma, hence its name. It is an aromatic spice with a taste similar to a combination of cinnamon, cloves and nutmeg, but hotter and more peppery.

History

Christopher Columbus discovered allspice in the Caribbean. Although he was seeking pepper, he had never actually seen real pepper and he thought allspice was it. He brought it back to Spain, where it got the name "pimienta," which is Spanish for pepper. Its Anglicized name, pimento, is occasionally used in the spice trade today. Before World War II, allspice was more widely used than it is nowadays. During the war, many trees producing allspice were cut, and production never fully recovered. Most allspice is produced in Jamaica, but some other sources for allspice include Guatemala, Honduras, as well as Mexico. Jamaican allspice is considered to be superior due to its higher oil content, which gives it a more appealing flavour.

Preparation/Form

Allspice is not, as is mistakenly believed by some people who have only come across it in ground form, a mixture of spices. Rather, it is the dried fruit of the *Pimenta dioica* plant. The fruit is picked when it is green and unripe, traditionally they are then sun dried. When dry they are brown and look like large brown peppercorns.

Allspice is most commonly sold as whole dried fruits or as a powder. The whole fruits have a longer shelf-life than the powdered product and produce a more aromatic product when freshly ground before use. Fresh leaves are also used where available: they are similar in texture to bay leaves and are thus infused during cooking and then removed before serving. Unlike bay leaves, they lose much flavour when dried and stored. The leaves and wood are often used for smoking meats where allspice is a local crop.

Uses

Allspice is one of the most important ingredients of Caribbean cuisine. It is used in Caribbean jerk seasoning (the wood is used to smoke jerk in Jamaica, although the spice is a good substitute), in mole sauces, and in pickling; it is also an ingredient in commercial sausage preparations and curry powders. Allspice is also indispensable in Middle Eastern cuisine, particularly in the Levant where it is used to flavour a variety of stews and meat dishes. In Palestinian cuisine, for example, many main dishes call for allspice as the sole spice added for flavoring. Allspice is commonly used in Great Britain and appears in many dishes, including in cakes. Even in many countries where allspice is not very popular in the household, such as Germany, it is used in large amounts by commercial sausage makers. Allspice is also a main flavour used in barbeque sauces.

Allspice has also been used as a deodorant, 18th century Russian soldiers would put allspice in their boots.

Folklore suggests that allspice provides relief for digestive problems. Volatile oils found in the plant contain eugenol, a weak antimicrobial agent (Yaniv, Sohara et al. 2005).

Cultivation

Allspice is a small shrubby tree, quite similar to the bay laurel in size and form. It can be grown outdoors in the tropics and subtropics with normal garden soil and watering. Smaller plants can be killed by frost, although larger plants are more tolerant. It adapts well to container culture and can be kept as a houseplant or in a greenhouse. The plant is dioecious, hence male and female plants must be kept in proximity in order to allow fruits to develop.

To protect the pimento trade the plant was guarded against export from Jamaica. It is reported that many attempts were made at growing the pimento from seeds, all failed. At one time it was thought that the plant would grow nowhere else except in Jamaica where the plant was readily spread by birds. Experiments were then performed using the constituents of bird droppings, however these were also totally unsuccessful. Eventually it was realized that an elevated temperature, such as that found inside a bird's body, was essential for germinating the seeds.

Camphor

Camphor is a white transparent waxy crystalline solid with a strong penetrating pungent aromatic odour. It is a terpenoid with the chemical formula $C_{10}H_{16}O$. It is found in wood of the camphor laurel (*Cinnamonum camphora*), a large evergreen tree found in Asia (particularly in Borneo and Taiwan, hence its alternate name) and some other related trees in the laurel family, notably *Ocotea usambarensis*; it can also be synthetically produced from oil of turpentine. It is used for its scent, as an ingredient in cooking (mainly in India), as an embalming fluid, in religious ceremonies and for medicinal purposes. A major source of camphor in Asia is Camphor basil.

History

The word camphor derives from the French word *camphre*, itself from Medieval Latin *camfora*, from Arabic *kafur*, from Malay *kapur Barus* meaning "Barus chalk". In fact Malay traders from whom Indian and Middle East merchants would buy camphor called it *kapur*, "chalk" because of its white colour. Barus was the port on the western coast of the Indonesian island of Sumatra where foreign traders would call to buy camphor. In the Indian language Sanskrit, the word 'karpoor' is used to denote Camphore. A south-indian adaptation of this word, 'karpooram' has been used for camphor in many south-indian/dravidian languages (like Telugu, Tamil, Kannada and Malayalam). Camphor was first synthesized by Gustaf Komppa in 1903. Previously, some organic compounds (such as urea) had been synthesized in the laboratory as a proof of concept, but camphor was a scarce natural product with a worldwide demand. The synthesis was the first industrial total synthesis, when Komppa began industrial production in Tainionkoski, Finland, in 1907.

Norcamphor is a camphor derivative with the three methyl groups replaced by hydrogen. Other substances deriving from trees are sometimes wrongly sold as camphor. Camphor Trees are widely found in very deep jungles of Western Ghats of Tamil Nadu and Kerala states in South India.

Uses

Modern uses include as a plasticizer for cellulose nitrate, as a moth repellent, as an antimicrobial substance, in embalming, and in fireworks. Camphor crystals are also used to prevent damage to insect collections by other small insects. A form of anti-itch gel currently on the market uses camphor as its active ingredient. It is also used in medicine. Camphor is readily absorbed through the skin and produces a feeling of cooling similar to that of menthol and acts as slight local anesthetic and antimicrobial substance. Camphor is an active ingredient (along with menthol) in vapour-steam products, such as Vicks VapoRub, and it is effective as a cough suppressant. It may also be administered orally in small quantities (50 mg) for minor heart symptoms and fatigue.

In the 17th Century, it was used by Auenbrugger in the treatment of mania. To prevent camphor from evaporating, just add few Black Peppers into the container of Camphor. Cockroachs, Snakes and other poisonous insects won't come near the camphor as camphor's strong smell drives them away and the camphor is poisonous for insects. Camphor is also used in the Mahashiva ratri celebrations of Shiva, the Hindu god of destruction of evil. It's natural pitch substance burns cool without leaving an ash residue, which symbolizes the consciousness.

Culinary

Currently, Camphor is mostly used as a flavoring for sweets in Asia. In ancient and medieval Europe it was widely used as ingredient for sweets but it is now mainly used for medicinal purposes. It is thought that camphor was used as a flavouring in confections resembling ice cream in China during the Tang dynasty (A.D. 618-907). Camphor is widely used in cooking (mainly for desert dishes) in India where it is known as *Pachha Karpooram* (literally meaning "Raw camphor" though "Pachha" means "Green" in Tamil). It is widely available at Indian grocery stores and is labelled as "Edible Camphor." In Hindu poojas and ceremonies, camphor is burned in a ceremonial spoon for performing aarti. This type of camphor is also sold at Indian grocery stores but it is not suitable for cooking. The only type that should be used for food are those which are labelled as "Edible Camphor."

Toxicology

In larger quantities, it is poisonous when ingested and can cause seizures, confusion, irritability, and neuromuscular hyperactivity. In 1980, the United States Food and Drug Administration set a limit of 11% allowable camphor in consumer products and totally banned products labelled as camphorated oil, camphor oil, camphor liniment, and camphorated liniment (but "white camphor essential oil" contains no significant amount of camphor). Since alternative treatments exist, medicinal use of camphor is discouraged by the FDA, except for skin-related uses, such as medicated powders, which contain only small amounts of camphor.

Nigella Sativa

Nigella sativa is an annual flowering plant, native to southwest Asia. It grows to 20-30 cm tall, with finely divided, linear (but not thread-like) leaves. The flowers are delicate, and usually coloured pale blue and white, with 5-10 petals. The fruit is a large and inflated capsule composed of 3-7 united follicles, each containing numerous seeds. The seed is used as a spice.

Nigella sativa seed is known variously as kalonji, kezah (Hebrew), charnushka (Russian), otu (Turkish), habbah Albarakah, (literally *seeds of blessing* Arabic) or siyah daneh (Persian). In English it is called fennel flower, black caraway, nutmeg flower, Roman coriander, or black onion seed. Other names used, sometimes misleadingly, are onion seed and black sesame (both of which are similar-looking but unrelated). Frequently the seeds are referred to as black cumin, this is, however, also used for a different spice, Bunium persicum. It is also sometimes just referred to as nigella or black seed. An old English name *gith* is now used for the corncockle.

This potpourri of vernacular names for this plant reflects that its widespread use as a spice is relatively new in the English speaking world, and largely associated with immigrants from areas where it is well known. Increasing use is likely to result in one of the names winning out, hopefully one which is unambiguous.

Nigella sativa has a pungent bitter taste and a faint smell of strawberries. It is used primarily in candies and liquors. The variety of naan bread called Peshawari naan is as a rule topped with kalonji seeds. In herbal medicine, *Nigella sativa* has hypertensive, carminative, and anthelminthic properties.

7

The Phyto-chemistry of Essential Oils

The Chemistry of Essential Oils, and their Chemical Components

Most people either use essential oils for their therapeutic effect or for the fragrance alone but it is also interesting to take note of the chemistry, of which the oils are made up from. It is interesting to know the chemical components that nature combines to make up the oils, but it is also humbling to take note of the fact that even with the best human efforts, should you in a laboratory combine all the chemicals in the correct proportions, you would still not have an identical oil. Such a copy of an oil will not have the same therapeutic effect as the natural and pure essential oil. And though we pride ourselves on being a technology advanced society, modern science can still not unlock the secrets of essential oils and why they can do what they do. If you for instance took all the correct chemical components, which will include lavandulol, borneol, terpineol, geraniol and linalol, and try to make up lavender essential oil in a laboratory, you will not have an oil that can successfully treat burns the way that true lavender oil can.

Essential oils, like all organic compounds, are made up of hydrocarbon molecules and can further be classified as terpenes, alcohols, esters, aldehydes, ketones and phenols etc. To help you in the understanding of the chemical constituents of the oils, it may be a good idea just to have a look at what an isoprene unit is. Every single oil normally has more than a hundred components, but this figure can also run into thousands, depending on the oil in question.

When you analyse essential oils with a chromatograph various organic components are found and the primary ones are as follows:

- Terpene hydrocarbons
  - o Monoterpene hydrocarbons
  - o Sesquiterpenes
- Oxygenated compounds
  - o Phenols
  - o Alcohols
    - * Monoterpene alcohols
    - * Sesquiterpene alcohols
  - o Aldehydes
  - o Ketones
  - o Esters
  - o Lactones
  - o Coumarins
  - o Ethers
  - o Oxides.

Terpenes Hydrocarbons

Monoterpene:

- These monoterpene compounds are found in nearly all essential oils and have a structure of 10 carbon atoms and at least one double bond. The 10 carbon atoms are derived from two isoprene units.
- They react readily to air and heat sources. For this reason citrus oils do not last well, since they are high in monoterpene hydrocarbons and have a quick reaction to air, and are readily oxidised.
- Although some quarters may simply state that these components have anti-inflammatory, antiseptic, antiviral and antibacterial therapeutic properties while some can be analgesic or stimulating with a tonic effect, it could be seen as a very broad generalisation, since this large group of chemicals vary greatly. Since some have a stimulating effect on the mucus membranes they are also often used as decongestants.

Sesquiterpenes:

- These sesquiterpenes consist of 15 carbon atoms and have complex pharmacological actions and here we can look at chamazulene, which is found in German chamomile.

- It has anti-inflammatory and anti-allergy properties. Another sesquiterpene often found in chamomile and rose, as well as other floral oils is farnesene.
- History highlight of terpene research
- The 1910 Nobel prize winner for Chemistry was Professor Otto Wallach for his work on terpenes which influenced the essential oil industry.

Oxygenated compounds:

- Phenols
  - o The phenols found in essential oils normally have a carbon side chain and here we can look at compounds such as thymol, eugenol and carvacrol. These components have great antiseptic, anti-bacterial and disinfectant qualities and also have greatly stimulating therapeutic properties.
  - o Due to the nature of phenols, essential oils that are high in them should be used in low concentrations and for short periods of time, since they can lead to toxicity if used over long periods of time, as the liver will be required to work harder to excrete them.
  - o Phenols are also classified as skin and mucus membrane irritants and although they have great antiseptic qualities, like cinnamon and clove oil, they can cause severe skin reactions.
- Alcohols
  - o Monoterpene alcohols
  - o These oils have good antiseptic, anti-viral and anti-fungal properties with very few side effects such as skin irritation or toxicity and have an uplifting energising effect.
  - o Examples of these alcohols are linalool, citronellol and terpineol found respectively in lavender, rose and geranium, and in juniper and tea tree oil.
  - o Sesquiterpene alcohols
  - o These alcohols are not commonly found in essential oils, but when found, like bisabolol in German chamomile, have great properties, which include liver and glandular stimulant, anti-allergen and anti-inflammatory.
  - o Other oils that contain sesquiterpene alcohols are sandalwood (a-santalol) as well as ginger, patchouli, vetiver, carrot seed, everlasting and valerian.

- Aldehydes
  - o These aldehydes have anti-fungal, anti-inflammatory, disinfectant, sedative yet uplifting therapeutic qualities and are the component that imparts the citrus-like fragrance in melissa, lemongrass and citronella. These properties are best used in aromatherapy when the essential oil is used in low dilutions - around 1%.
  - o Should oils high in this component be used, it could cause skin irritation and sensitivity as for instance lemongrass oil. Aldehydes are also unstable and will easily oxidise in the presence of oxygen and even low heat.
- Ketones
  - o Although ketones can be toxic, as in the case of thujone found in thuja and wormwood oil as well as pinocamphone found in others, they also have some great therapeutic benefits - especially in the field of easing the secretion of mucus as well as cell and tissue regeneration.
  - o Other oils, such as hyssop, eucalyptus and rosemary have moderate amounts of ketones, and when used properly in aromatherapy can be greatly beneficial to the body.
  - o The ketone italidone found in everlasting, not only has the mucolytic (mucus easing) properties, but is also useful in skin regeneration, wound healing and reducing old scar tissue such as in wounds, stretch marks and adhesions.
  - o Essential oils high in ketones need to be used with care in pregnancy.
- Esters
  - o Esters are formed from alcohols and acids, and are named after both their original molecules with the alcohols dropping the "ol" and gaining an "yl" and the acids dropping the "ic" and gaining an "ate".
  - o The esters found in essential oils are normally very fragrant and tend to be fruity and their therapeutic effects include being sedative and antispasmodic. Some esters also have anti-fungal and anti-microbial properties - like the anti-fungal properties in geranium oil.
  - o The most well known ester must be lineally acetate, which is found in lavender, clary sage as well as petitgrain.
  - o These components are normally gentle in their actions and can be used with great ease.

- Lactones and coumarins
  - Lactones contain an ester group integrated into a carbon ring system and coumarins are also types of lactones. There are similarities between the actions of lactones, coumarins and ketones since they also have some neurotoxic effects and can cause skin sensitising and irritation.
  - Yet the sesquiterpene lactone, called helenalin found in arnica oil, seems to be responsible for the anti-inflammatory action of arnica oil.
  - The amount of lactones and coumarins normally found in essential oils is very low, and does not pose a huge problem. Lactones also have great mucus moving and expectorant properties and for this reason elecampane is often used in the treatment of bronchitis and chest complaints.
  - Some coumarins, like furocoumarin - bergaptene - found in bergamot oil are severely skin UV sensitive and should be used with great care should you be exposed to sunlight.
- Ethers
  - Phenolic ethers are the most widely found ethers in essential oils with anethole found in aniseed, the only real ether of importance together with methyl chavicol found in basil and tarragon.
- Oxides
  - The main therapeutic effect of oxides are that of expectorant, with 1,8-cineole - commonly known as eucalyptol being the most well known.

8

Edible Oils

Major Oils

These oils account for a significant fraction of worldwide edible oil production. All are also used as fuel oils.

Figure : *Sunflowers, the seeds of which are the source of Sunflower oil.*

- Coconut oil, a cooking oil, with medical and industrial applications as well. Extracted from the kernel or meat of the fruit of the coconut palm. Common in the tropics, and unusual in composition, with medium chain fatty acids dominant.
- Corn oil, one of the principal oils sold as salad and cooking oil.
- Cottonseed oil, used as a salad and cooking oil, both domestically and industrially.
- Olive oil, used in cooking, cosmetics, soaps, and as a fuel for traditional oil lamps.
- Palm oil, the most widely produced tropical oil. Popular in West African and Brazilian cuisine. Also used to make biofuel.
- Peanut oil (Ground nut oil), a clear oil with some applications as a salad dressing, and, due to its high smoke point, especially used for frying.

- Rapeseed oil, including Canola oil, one of the most widely used cooking oils.
- Safflower oil, until the 1960s used in the paint industry, now mostly as a cooking oil.
- Sesame oil, cold pressed as light cooking oil, hot pressed for a darker and stronger flavour.
- Soybean oil, produced as a byproduct of processing soy meal.
- Sunflower oil, a common cooking oil, also used to make biodiesel.

Nut Oils

Figure : *Hazelnuts from the Common Hazel, used to make Hazelnut oil.*

Nut oils are generally used in cooking, for their flavour. Most are quite costly, because of the difficulty of extracting the oil.

- Almond oil, used as an edible oil, but primarily in the manufacture of cosmetics.
- Beech nut oil, from *Fagus sylvatica* nuts, is a well-regarded edible oil in Europe, used for salads and cooking.
- Cashew oil, somewhat comparable to olive oil. May have value for fighting dental cavities.
- Hazelnut oil, mainly used for its flavour. Also used in skin care, because of its slight astringent nature.
- Macadamia oil, with a mild nutty flavour and a high smoke point.
- Mongongo nut oil (or *manketti oil*), from the seeds of the *Schinziophyton rautanenii*, a tree which grows in South Africa. High in vitamin E. Also used in skin care.
- Pecan oil, valued as a food oil, but requiring fresh pecans for good quality oil.
- Pine nut oil, sold as a gourmet cooking oil, and of potential medicinal interest as an appetite suppressant.

- Pistachio oil, a strongly flavoured oil with a distinctive green colour.
- Walnut oil, used for its flavour, also used by Renaissance painters in oil paints.

Citrus Oils

A number of citrus plants yield pressed oils. Some, like lemon and orange oil, are used as essential oils, which is uncommon for pressed oils. The seeds of many if not most members of the citrus family yield usable oils.

- Grapefruit seed oil, extracted from the seeds of grapefruit (*Citrus* × *paradisi*). Grapefruit seed oil was extracted experimentally in 1930 and was shown to be suitable for making soap.
- Lemon oil, similar in fragrance to the fruit. One of a small number of cold pressed essential oils. Used as a flavouring agent and in aromatherapy.
- Orange oil, like lemon oil, cold pressed rather than distilled. Consists of 90% d-Limonene. Used as a fragrance, in cleaning products and in flavouring foods.

Oils from Melon and Gourd Seeds

Figure : *Watermelon seed oil, extracted from the seeds of* Citrullus vulgaris, *is used in cooking in West Africa.*

Members of the Cucurbitaceae include gourds, melons, pumpkins, and squashes. Seeds from these plants are noted for their oil content, but little information is available on methods of extracting the oil. In most cases, the plants are grown as food, with dietary use of the oils as a byproduct of using the seeds as food.

- Bitter gourd oil, from the seeds of *Momordica charantia.* High in α-Eleostearic acid. Of current research interest for its potential anti-carcinogenic properties.
- Bottle gourd oil, extracted from the seeds of the *Lagenaria siceraria,* widely grown in tropical regions. Used as an edible oil.
- Buffalo gourd oil, from the seeds of the *Cucurbita foetidissima,* a vine with a rank odor, native to southwest North America.
- Butternut squash seed oil, from the seeds of *Cucurbita moschata,* has a nutty flavour that is used for salad dressings, marinades, and sautéeing.
- Egusi seed oil, from the seeds of *Cucumeropsis mannii naudin,* is particularly rich in linoleic acid.
- Pumpkin seed oil, a speciality cooking oil, produced in Austria, Slovenia and Croatia. Used mostly in salad dressings.
- Watermelon seed oil, pressed from the seeds of *Citrullus vulgaris.* Traditionally used in cooking in West Africa.

Food Supplements

A number of oils are used as food supplements (or "nutraceuticals"), for their nutrient content or purported medicinal effect. Borage seed oil, blackcurrant seed oil, and evening primrose oil all have a significant amount of gamma-Linolenic acid (GLA) (about 23%, 15–20% and 7–10%, respectively), and it is this that has drawn the interest of researchers.

- Açaí oil, from the fruit of several species of the Açaí palm (*Euterpe*) grown in the Amazon region.
- Black seed oil, pressed from *Nigella sativa* seeds, has a long history of medicinal use, including in ancient Greek, Asian, and Islamic medicine, as well as a topic of current medical research.
- Blackcurrant seed oil, from the seeds of *Ribes nigrum,* used as a food supplement. High in gamma-Linolenic, omega-3 and omega-6 fatty acids.
- Borage seed oil, from the seeds of *Borago officinalis,* with an omega-3 content comparable to blackcurrant seed oil and evening primrose oil.
- Evening primrose oil, from the seeds of *Oenothera biennis,* the most important plant source of gamma-Linolenic acid, particularly because it does not contain alpha-Linolenic acid.

- Flaxseed oil (called linseed oil when used as a drying oil), from the seeds of *Linum usitatissimum*. High in omega-3 and lignans, which can be used medicinally. A good dietary equivalent to fish oil. Easily turns rancid.

Other Edible Oils

Figure : *Carob seed pods, used to make carob pod oil.*

- Amaranth oil, from the seeds of grain amaranth species, including *Amaranthus cruentus* and *Amaranthus hypochondriacus*, high in squalene and unsaturated fatty acids.
- Apricot oil, similar to almond oil, which it resembles. Used in cosmetics.
- Apple seed oil, high in linoleic acid.
- Argan oil, from the seeds of the *Argania spinosa*, is a food oil from Morocco developed through a women's cooperative founded in the 1990s, that has also attracted recent attention in Europe.
- Avocado oil, an edible oil used primarily in the cosmetics and pharmaceutical industries. Unusually high smoke point of 510°F.
- Babassu oil, from the seeds of the *Attalea speciosa*, is similar to, and used as a substitute for, coconut oil.
- Ben oil, extracted from the seeds of the *Moringa oleifera*. High in behenic acid. Extremely stable edible oil. Also suitable for biofuel.
- Borneo tallow nut oil, extracted from the fruit of species of genus *Shorea*. Used as a substitute for cocoa butter, and to make soap, candles, cosmetics and medicines in places where the tree is common.
- Cape chestnut oil, also called yangu oil, is a popular oil in Africa for skin care.

- Carob pod oil (Algaroba oil), from carob, with an exceptionally high essential fatty acid content.
- Cocoa butter, from the cacao plant. Used in the manufacture of chocolate, as well as in some cosmetics.
- Cocklebur oil, from species of genus *Xanthium*, with similar properties to poppyseed oil, similar in taste and smell to sunflower oil.
- Cohune oil, from the *Attalea cohune* (cohune palm) used as a lubricant, for cooking, soapmaking and as a lamp oil.
- Coriander seed oil, from coriander seeds, used in a wide variety of flavouring applications, including gin and seasoning blends. Recent research has shown promise for use in killing food-borne bacteria, such as *E. coli.*
- Date seed oil, extracted from date pits. Its low extraction rate and lack of other distinguishing characteristics make it an unlikely candidate for major use.
- Dika oil, from *Irvingia gabonensis* seeds, native to West Africa. Used to make margarine, soap and pharmaceuticals, where is it being examined as a tablet lubricant. Largely underdeveloped.
- False flax oil made of the seeds of *Camelina sativa.* One of the earliest oil crops, dating back to the 6th millennium B.C. Produced in modern times in Central and Eastern Europe; fell out of production in the 1940s. Considered promising as a food or fuel oil.
- Grape seed oil, a cooking and salad oil, also sprayed on raisins to help them retain their flavour.
- Hemp oil, a high quality food oil also used to make paints, varnishes, resins and soft soaps.
- Kapok seed oil, from the seeds of *Ceiba pentandra*, used as an edible oil, and in soap production.
- Kenaf seed oil, from the seeds of *Hibiscus cannabinus.* An edible oil similar to cottonseed oil, with a long history of use.
- Lallemantia oil, from the seeds of *Lallemantia iberica*, discovered at archaeological sites in northern Greece.
- Mafura oil, extracted from the seeds of *Trichilia emetica.* Used as an edible oil in Ethiopia. Mafura butter, extracted as part of the same process when extracting the oil, is not edible, and is used in soap and candle making, as a body ointment, as fuel, and medicinally.

- Marula oil, extracted from the kernel of *Sclerocarya birrea*. Used as an edible oil with a light, nutty flavour. Also used in soaps. Fatty acid composition is similar to that of olive oil.
- Meadowfoam seed oil, highly stable oil, with over 98% long-chain fatty acids. Competes with rapeseed oil for industrial applications.
- Mustard oil (pressed), used in India as a cooking oil. Also used as a massage oil.
- Nutmeg butter, extracted by expression from the fruit of cogeners of genus *Myristica*. Nutmeg butter has a large amount of trimyristin. Nutmeg oil, by contrast, is an essential oil, extracted by steam distillation.
- Okra seed oil, from *Abelmoschus esculentus*. Composed predominantly of oleic and linoleic acids. The greenish yellow edible oil has a pleasant taste and odor.
- Papaya seed oil, high in omega-3 and omega-6, similar in composition to olive oil. Not to be confused with papaya oil produced by maceration.
- Perilla seed oil, high in omega-3 fatty acids. Used as an edible oil, for medicinal purposes in Asian herbal medicine, in skin care products and as a drying oil.
- Persimmon seed oil, extracted from the seeds of *Diospyros virginiana*. Dark, reddish brown colour, similar in taste to olive oil. Nearly equal content of oleic and linoleic acids.
- Pequi oil, extracted from the seeds of *Caryocar brasiliense*. Used in Brazil as a highly prized cooking oil.
- Pili nut oil, extracted from the seeds of *Canarium ovatum*. Used in the Philippines as an edible oil, as well as for a lamp oil.
- Pomegranate seed oil, from *Punica granatum* seeds, is very high in punicic acid (which takes its name from pomegranates). A topic of current medical research for treating and preventing cancer.
- Poppyseed oil, long used for cooking, in paints, varnishes, and soaps.
- Prune kernel oil, marketed as a gourmet cooking oil Similar in composition to peach kernel oil.
- Quinoa oil, similar in composition and use to corn oil.
- Ramtil oil, pressed from the seeds of the one of several species of genus *Guizotia abyssinica* (Niger pea) in India and Ethiopia.

- Rice bran oil is a highly stable cooking and salad oil, suitable for high-temperature cooking. It also has potential as a biofuel.
- Royle oil, pressed from the seeds of *Prinsepia utilis*, a wild, edible oil shrub that grows in the higher Himalayas. Used medicinally in Nepal.
- Sacha inchi oil, from the Peruvian Amazon. High in behenic, omega-3 and omega-6 fatty acids.
- Sapote oil, used as a cooking oil in Guatemala.
- Seje oil, from the seeds of *Jessenia bataua*. Used in South America as an edible oil, similar to olive oil, as well as for soaps and in the cosmetics industry.
- Shea butter, much of which is produced by poor, African women. Used primarily in skin care products and as a substitute for cocoa butter in confections and cosmetics.
- Taramira oil, from the seeds of the arugula (*Eruca sativa*), grown in West Asia and Northern India. Used as a (pungent) edible oil after aging to remove acridity.
- Tea seed oil (Camellia oil), widely used in southern China as a cooking oil. Also used in making soaps, hair oils and a variety of other products.
- Thistle oil, pressed from the seeds of *Silybum marianum*. A good potential source of special fatty acids, carotenoids, tocopherols, phenol compounds and natural anti-oxidants, as well as for generally improving the nutritional value of foods.
- Tigernut oil (or nut-sedge oil) is pressed from the tuber of *Cyperus esculentus*. It has properties similar to soybean, sunflower and rapeseed oils. It is used in cooking and making soap and has potential as a biodiesel fuel.
- Tobacco seed oil, from the seeds of *Nicotiana tabacum* and other *Nicotiana* species. If purified, is suitable for edible purposes.
- Tomato seed oil is a potentially valuable by-product, as a cooking oil, from the waste seeds generated from processing tomatoes.
- Wheat germ oil, used nutritionally and in cosmetic preparations, high in vitamin E and octacosanol.

Oils Used for Biofuel

A number of oils are used for biofuel (biodiesel and Straight Vegetable Oil) in addition to having other uses. Other oils are used only as biofuel.

Although diesel engines were invented, in part, with vegetable oil in mind, diesel fuel is almost exclusively petroleum-based. Vegetable oils are evaluated for use as a biofuel based on:

1. Suitability as a fuel, based on flash point, energy content, viscosity, combustion products and other factors
2. Cost, based in part on yield, effort required to grow and harvest, and post-harvest processing cost

Figure : *A flask of biodiesel*

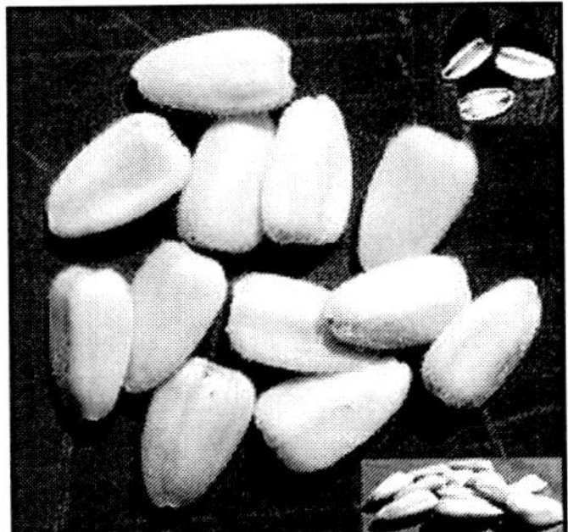

Figure : *Sunflower kernels*

Figure : *Jojoba fruit*

Multipurpose Oils also Used as Biofuel

The oils listed immediately below are all (primarily) used for other purposes – all but tung oil are edible – but have been considered for use as biofuel.

- Castor oil, lower cost than many candidates. Kinematic viscosity may be an issue.
- Coconut oil (copra oil), promising for local use in places that produce coconuts.
- Colza oil, from *Brassica rapa, var. oleifera* (turnip) is closely related to rapeseed (or canola) oil. It is a major source of biodiesel in Germany.
- Corn oil, appealing because of the abundance of maize as a crop.
- Cottonseed oil, the subject of study for cost-effectiveness as a biodiesel feedstock.
- False flax oil, from *Camelina sativa*, used in Europe in oil lamps until the 18th century.
- Hemp oil, relatively low in emissions. Production is problematic in some countries because of its association with marijuana.
- Mustard oil, shown to be comparable to Canola oil as a biofuel.
- Palm oil, very popular for biofuel, but the environmental impact from growing large quantities of oil palms has recently called the use of palm oil into question.
- Peanut oil, used in one of the first demonstrations of the Diesel engine in 1900.
- Radish oil. Wild radish contains up to 48% oil, making it appealing as a fuel.
- Rapeseed oil, the most common base oil used in Europe in biodiesel production.
- Ramtil oil, used for lighting in India.
- Rice bran oil, appealing because of lower cost than many other vegetable oils. Widely grown in Asia.
- Safflower oil, explored recently as a biofuel in Montana.
- Salicornia oil, from the seeds of *Salicornia bigelovii*, a halophyte (salt-loving plant) native to Mexico.
- Soybean oil, not economical as a fuel crop, but appealing as a byproduct of soybean crops for other uses.
- Sunflower oil, suitable as a fuel, but not necessarily cost effective.

- Tigernut oil has been described by researchers in China as having "great potential as a biodiesel fuel."
- Tung oil, referenced in several lists of vegetable oils that are suitable for biodiesel. Several factors in China produce biodiesel from tung oil.

Inedible Oils Used only or Primarily as Biofuel

These oils are extracted from plants that are cultivated solely for producing oil-based biofuel. These, plus the major oils described above, have received much more attention as fuel oils than other plant oils.

- Copaiba, an oleoresin tapped from species of genus *Copaifera*. Used in Brazil as a cosmetic product and a major source of biodiesel.
- Honge oil (Pongamia), pioneered as a biofuel by Udipi Shrinivasa in Bangalore, India.
- Jatropha oil, widely used in India as a fuel oil. Has attracted strong proponents for use as a biofuel.
- Jojoba oil, from the *Simmondsia chinensis*, a desert shrub.
- Milk bush, popularised by chemist Melvin Calvin in the 1950s. Researched in the 1980s by Petrobras, the Brazilian national petroleum company.
- Nahor oil, pressed from the kernels of *Mesua ferrea*, is used in India as a lamp oil.
- Paradise oil, from the seeds of *Simarouba glauca*, has received interest in India as a feed stock for biodiesel.
- Petroleum nut oil, from the Petroleum nut (*Pittosporum resiniferum*) native to the Philippines. The Philippine government once explored the use of the petroleum nut as a biofuel.

Drying Oils

Drying oils are vegetable oils that dry to a hard finish at normal room temperature. Such oils are used as the basis of oil paints, and in other paint and wood finishing applications. In addition to the oils listed here, walnut, sunflower and safflower oil are also considered to be drying oils.

- Dammar oil, from the *Canarium strictum*, used in paint as an oil drying agent. Can also be used as a lamp oil.
- Linseed oil's properties as a polymer make it highly suitable for wood finishing, for use in oil paints, as a plasticizer and

hardener in putty and in making linoleum. When used in food or medicinally, linseed oil is called flaxseed oil.

- Poppyseed oil, similar in usage to linseed oil but with better colour stability.
- Stillingia oil (also called *Chinese vegetable tallow oil*), obtained by solvent from the seeds of *Sapium sebiferum*. Used as a drying agent in paints and varnishes.
- Tung oil, used as an industrial lubricant and highly effective drying agent. Also used as a substitute for linseed oil.
- Vernonia oil is produced from the seeds of the Vernonia galamensis. It is composed of 73–80% vernolic acid, which can be used to make epoxies for manufacturing adhesives, varnishes and paints, and industrial coatings.

Other Oils

A number of pressed vegetable oils are either not edible, or not used as an edible oil.

- Amur cork tree fruit oil, pressed from the fruit of the *Phellodendron amurense*. It has been studied for insecticidal use.
- Artichoke oil, extracted from the seeds of the artichoke fruit, is an unsaturated semi-drying oil with potential applications in making soap, shampoo, alkyd resin and shoe polish.
- Balanos oil, pressed from the seeds of *Balanites aegyptiaca*, was used in ancient Egypt as the base for perfumes.
- Bladderpod oil, pressed from the seeds of *Lesquerella fendleri*, native to North America. Rich in lesquerolic acid, which is chemically similar to the ricinoleic acid found in castor oil. Many industrial uses. Possible substitute for castor oil as it requires much less moisture than castor beans.

Figure : *The fruit of the amur cork tree*

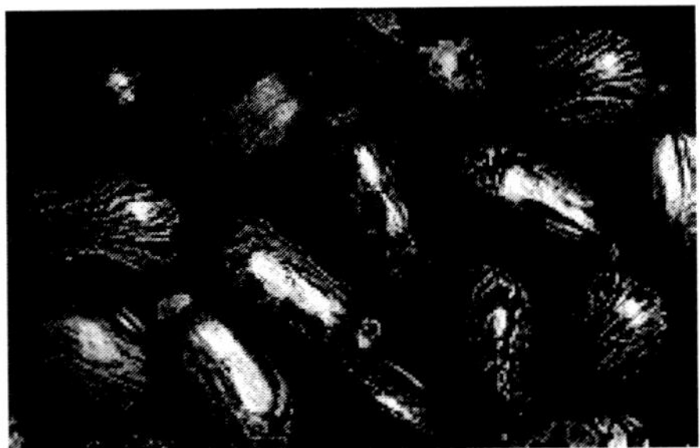

Figure : *Castor beans are the source of castor oil*

- Brucea javanica oil, extracted from the seeds of the *Brucea javanica*. The oil has been shown to be effective in treating certain cancers.
- Burdock oil (Bur oil) extracted from the root of the burdock. Used as an herbal remedy for scalp conditions.
- Candlenut oil (Kukui nut oil), produced in Hawai'i, used primarily for skin care products.
- Carrot seed oil (pressed), from carrot seeds, used in skin care products.
- Castor oil, with many industrial and medicinal uses. Castor beans are also a source of the toxin ricin.
- Chaulmoogra oil, from the seeds of *Hydnocarpus wightiana*, used for many centuries, internally and externally, to treat leprosy. Also used to treat secondary syphilis, rheumatism, scrofula, and in phthisis.
- Crambe oil, extracted from the seeds of the *Crambe abyssinica*. High in erucic acid, used as an industrial lubricant, a corrosion inhibitor, and as an ingredient in the manufacture of synthetic rubber.
- Croton oil (tiglium oil) is pressed from the seeds of *Croton tiglium*. Highly toxic, it was formerly used as a drastic purgative.
- Cuphea oil, from a number of species of genre *Cuphea*. Of interest as sources of medium chain triglycerides.
- Honesty oil, from the seeds of *Lunaria annua*, which contain 30–40% oil. The oil is particularly rich in long chain fatty acids, including erucic and nervonic acid, making it suitable for certain industrial purposes.
- Illipe butter, from the nuts of the *Shorea stenoptera*. Similar to cocoa butter, but with a higher melting point. Used in cosmetics.

- Jojoba oil, used in cosmetics as an alternative to whale oil spermaceti.
- Mango oil, pressed from the stones of the mango fruit, is high in stearic acid, and can be used for making soap.
- Mowrah butter, from the seeds of the *Madhuca latifolia* and *Madhuca longifolia*, both native to India. Crude Mowrah butter is used as a fat for spinning wool, for making candles and soap. The refined fat is used as an edible fat and vegetable ghee in India.
- Neem oil, from *Azadirachta indica*, a brownish-green oil with a high sulphur content, used in cosmetics, for medicinal purposes, and as an insecticide.
- Ojon oil extracted from the nut of the American palm (*Elaeis oleifera*). Oil extracted from both the nut and husk is also used as an edible oil in Central and South America. Commercialised by a Canadian businessman in the 1990s.
- Rose hip seed oil, used primarily in skin care products, particularly for aging or damaged skin.
- Rubber seed oil, pressed from the seeds of the Rubber tree (*Hevea brasiliensis*), has received attention as a potential use of what otherwise would be a waste product from making rubber. It has been explored as a drying oil in Nigeria, as a diesel fuel in India and as food for livestock in Cambodia and Vietnam.
- Sea buckthorn oil, derived from *Hippophae rhamnoides*, produced in northern China, used primarily medicinally.
- Sea rocket seed oil, from the halophyte *Cakile maritima*, native to north Africa, is high in erucic acid, and therefore has potential industrial applications.
- Snowball seed oil (Viburnum oil), from *Viburnum opulus* seeds. High in tocopherol, carotenoides and unsaturated fatty acids. Used medicinally.
- Tall oil, produced as a byproduct of wood pulp manufacture. A further byproduct called *tall oil fatty acid* (TOFA) is a cheap source of oleic acid.
- Tamanu or foraha oil from the *Calophyllum tacamahaca*, is important in Polynesian culture, and, although very expensive, is used for skin care.
- Tonka bean oil (Cumaru oil), used medicinally in Brazil.
- Ucuhuba seed oil, extracted from the seeds of *Virola surinamensis*, is unusually high in myristic acid.

Bioliquids

Bioliquids are liquid fuels made from biomass for energy purposes other than transport (i.e. heating and electricity). Bioliquids are usually made from virgin or used vegetable and seed oils, like palm or soya oil. These oils are burned in a power station to create heat, which can then be used to warm homes or boil water to make steam. This steam can then be used to drive a turbine to generate electricity. Rudolf Diesel's first public exhibition of the internal combustion engine, that was to later bear his name, ran on peanut oil.

Bioliquid Production and Use

Bioliquids have been used for many years to provide heat for homes on a small scale but now big energy providers are looking at their use on a larger scale. A controversial plant in Bristol (UK) was recently given the go ahead despite receiving several hundred complaints. The plant will be built and operated by W4B and provide enough power for 25,000 homes.

Advantages

Bioliquids have several key advantages over other sources of renewable energy:

- Bioliquids have a high energy density
- The technology is well established, having been used for many years
- Can be used on demand, reacting quickly to changes in demand for power.

Disadvantages

Many of the same problems that affect biofuels also affect bioliquids and there are various social, economic, environmental and technical issues, which have been discussed in the popular media and scientific journals. These include: the effect of moderating oil prices, the "food vs fuel" debate, poverty reduction potential, carbon emissions levels, sustainable biofuel production, deforestation and soil erosion, loss of biodiversity, impact on water resources, as well as energy balance and efficiency.

Bioliquids also have several key problems compared to other sources of renewable energy:

- Price of fuel is very variable, due to competitiveness of feedstock for other uses (e.g. soap)

- Supply chain is still very new
- Governments, such as the EU, remained undecided on bioliquids

Vegetable Oil Fuel

Vegetable oil is an alternative fuel for diesel engines and for heating oil burners. For engines designed to burn diesel fuel, the viscosity of vegetable oil must be lowered to allow for proper atomisation of the fuel, otherwise incomplete combustion and carbon build up will ultimately damage the engine. Many enthusiasts refer to vegetable oil used as fuel as waste vegetable oil (WVO) if it is oil that was discarded from a restaurant or straight vegetable oil (SVO) or pure plant oil (PPO) to distinguish it from biodiesel.

Rudolf Diesel was the father of the engine which bears his name. His first attempts were to design an engine to run on coal dust, but later designed his engine to run on vegetable oil. The idea, he hoped, would make his engines more attractive to farmers having a source of fuel readily available. In a 1912 presentation to the British Institute of Mechanical Engineers, he cited a number of efforts in this area and remarked, "The fact that fat oils from vegetable sources can be used may seem insignificant today, but such oils may perhaps become in course of time of the same importance as some natural mineral oils and the tar products are now."

Periodic petroleum shortages spurred research into vegetable oil as a diesel substitute during the 1930s and 1940s, and again in the 1970s and early 1980s when straight vegetable oil enjoyed its highest level of scientific interest. The 1970s also saw the formation of the first commercial enterprise to allow consumers to run straight vegetable oil in their automobiles, Elsbett of Germany. In the 1990s Bougainville conflict, islanders cut off from oil supplies due to a blockade used coconut oil to fuel their vehicles.

Academic research into straight vegetable oil fell off sharply in the 1980s with falling petroleum prices and greater interest in biodiesel as an option that did not require extensive vehicle modification.

Application and Usability

While engineers and enthusiasts have been experimenting with using vegetable oils as fuel for a diesel engine since at least 1900, it is only recently that the necessary fuel properties and engine parameters for reliable operation have become apparent. Only a handful of peer reviewed studies exist that show reliable long term use of vegetable oil.

Modified Fuel Systems

Most diesel car engines are suitable for the use of SVO, also commonly called pure plant oil (PPO), with suitable modifications. Principally, the viscosity and surface tension of the SVO/PPO must be reduced by preheating it, typically by using waste heat from the engine or electricity, otherwise poor atomisation, incomplete combustion and carbonisation may result. One common solution is to add a heat exchanger, and an additional fuel tank for "normal" diesel fuel (petrodiesel or biodiesel) and a three way valve to switch between this additional tank and the main tank of SVO/PPO. (This aftermarket modification typically costs about $1,200 USD.) The engine is started on diesel, switched over to vegetable oil as soon as it is warmed up and switched back to diesel shortly before being switched off to ensure that no vegetable oil remains in the engine or fuel lines when it is started from cold again. In colder climates it is often necessary to heat the vegetable oil fuel lines and tank as it can become very viscous and even solidify.

Single tank conversions have been developed, largely in Germany, which have been used throughout Europe. These conversions are designed to provide reliable operation with rapeseed oil that meets the German rapeseed oil fuel standard DIN 51605. Modifications to the engines cold start regime assist combustion on start up and during the engine warm up phase. Suitably modified indirect injection (IDI) engines have proven to be operable with 100% PPO down to temperatures of "10°C. Direct injection (DI) engines generally have to be preheated with a block heater or diesel fired heater. The exception is the VW Tdi (Turbocharged Direct Injection) engine for which a number of German companies offer single tank conversions. For long term durability it has been found necessary to increase the oil change frequency and to pay increased attention to engine maintenance.

Unmodified Indirect Injection Engines

Many cars powered by indirect injection engines supplied by in-line injection pumps, or mechanical Bosch injection pumps are capable of running on pure SVO/PPO in all but winter temperatures. Indirect injection Mercedes-Benz vehicles with in-line injection pumps and cars featuring the PSA XUD engine tend to perform reasonably, especially as the latter is normally equipped with a coolant heated fuel filter. Engine reliability would depend on the condition of the engine. Attention to maintenance of the engine, particularly of the fuel injectors, cooling system and glow plugs will help to provide longevity. Ideally the engine would be converted.

Vegetable Oil Blending

The relatively high kinematic viscosity of vegetable oils must be reduced to make them compatible with conventional compression-ignition engines and fuel systems. Cosolvent blending is a low-cost and easy-to-adapt technology that reduces viscosity by diluting the vegetable oil with a low-(molecular weight solvent). This blending, or "cutting", has been done with diesel fuel, kerosene, and gasoline, amongst others; however, opinions vary as to the efficacy of this. Noted problems include higher rates of wear and failure in fuel pumps and piston rings when using blends.

Home Heating

When liquid fuels made from biomass are used for energy purposes other than transport, they are called bioliquids. With often minimal modification, most residential furnaces and boilers that are designed to burn No. 2 heating oil can be made to burn either biodiesel or filtered, preheated waste vegetable oil. These are generally not as clean-burning as petroleum fuel oil, but if processed at home, by the consumer, can result in considerable savings. Many restaurants will give away their used cooking oil either free or at minimal cost, and processing to biodiesel is fairly simple and inexpensive. Burning filtered waste vegetable oil (WVO) directly is somewhat more problematic, since it is much more viscous, but it can be accomplished with suitable preheating. WVO can thus be an economical heating option for those with the necessary mechanical and experimental aptitude, and where fire regulations and insurance policy permit it.

Material Compatibility

Free fatty acids in WVO can have a detrimental effect on metals. Copper and its alloys, such as brass, are affected. Zinc and zinc-plating (galvanization) are stripped by FFA's and tin, lead, iron, and steel are affected too. Stainless steel and aluminum are generally unaffected.

Temperature Effects

Some Pacific island nations are using coconut oil as fuel to reduce their expenses and their dependence on imported fuels while helping stabilise the coconut oil market. Coconut oil is only usable where temperatures do not drop below 17 degrees Celsius (62 degrees Fahrenheit), unless two-tank SVO/PPO kits or other tank-heating accessories, etc. are used. Fortunately, the same techniques developed to use, for example, canola and other oils in cold climates can be implemented to make coconut oil usable in temperatures lower than 17 degrees Celsius.

Availability

Waste Vegetable Oil

As of 2000, the United States was producing in excess of 11 billion litres (2.9 billion U.S. gallons) of waste vegetable oil annually, mainly from industrial deep fryers in potato processing plants, snack food factories and fast food restaurants. If all those 11 billion litres could be collected and used to replace the energy equivalent amount of petroleum (an ideal case), almost 1% of US oil consumption could be offset. Use of waste vegetable oil as a fuel competes with some other uses of the commodity, which has effects on its price as a fuel and increases its cost as an input to the other uses as well.

Pure Plant Oil (Straight Vegetable Oil)

Pure plant oil (PPO) (or Straight Vegetable Oil (SVO)), in contrast to waste vegetable oil, is not a byproduct of other industries, and thus its prospects for use as fuel are not limited by the capacities of other industries. Production of vegetable oils for use as fuels is theoretically limited only by the agricultural capacity of a given economy. However, doing so detracts from the supply of other uses of pure vegetable oil.

Legal Implications

Taxation of Fuel

Taxation on SVO/PPO as a road fuel varies from country to country, and it is possible the revenue departments in many countries are even unaware of its use, or feel it too insignificant to legislate. Germany used to have 0% taxation, resulting in it being a leader in most developments of the fuel use. However SVO/PPO as a road fuel began to be taxed at 0,09 €/litre from 1 January 2008 in Germany, with incremental rises up to 0,45 €/litre by 2012. However, in Australia it has become illegal to produce any fuel if it is to be sold unless a license to do so is granted by the federal government. This is a chargeable offence with a fine of up to 20,000 dollars but this bracket may alter circumstantially. Also a jail term may result if offenders are aware of the illegality of selling the fuel.

The legality of burning SVO in the United States of America is debated by many. Though vehicle conversions are available both as "do-it-yourself" kits, and professionally installed in virtually every metropolitan area, the EPA clearly states vegetable oil (raw or recycled) is not registered for use as a vehicle fuel. Further, vehicles converted to use vegetable oil as fuel would "likely need to be certified by the EPA", and no such certifications have been done to date.

There seems to be no clear federal taxation system in the USA. Production of biodiesel in some US regions may require motor fuel taxes to be paid.

The Japanese Government has also exempted the use of SVO as a fuel from road tax.

In the Republic of Ireland a pilot scheme is currently running, whereby as of April 2006, eight suppliers have been approved to sell SVO/PPO for use as a fuel without the payment of excise duty (Value Added Tax at 21% still applies). SVO from any other source still attracts excise duty at 36.8058 Euro cents per litre plus 21% VAT).

Despite its prevalent use in France, it appears there has been no legislation to date to cover this. In the UK it is legal once duty on the fuel is paid, or, in the case of using or producing less than 2,500 litres per year, no duty is necessary. In the UK, drivers using SVO/PPO have in the past been prosecuted for failure to pay duty to Her Majesty's Revenue and Customs. The rate of taxation on SVO was originally set at a reduced rate of 27.1p per litre, but in late 2005, HMRC started to enforce the full diesel excise rate of 47.1p per litre.

HMRC argued that SVOs/PPOs on the market from small producers did not meet the official definition of "biodiesel" in Section 2AA of The Hydrocarbon Oil Duties Act 1979 (HODA), and consequently was merely a "fuel substitute" chargeable at the normal diesel rate. Such a policy seemed to contradict the UK Government's commitments to the Kyoto Protocol and to many EU directives and had many consequences, including an attempt to make the increase retroactive, with one organisation being presented with a £16,000 back tax bill. This change in the rate of excise duty effectively removed any commercial incentive to use SVO/PPO, regardless of its desirability on environmental grounds; unless waste vegetable oil can be obtained free of charge, the combined price of SVO/PPO and taxation for its use usually exceeded the price of mineral diesel. HMRC's interpretation is widely challenged by the SVO/PPO industry and the UK pure Plant Oil Association (UKPPOA) was formed to represent the interests of people using vegetable oil as fuel and to lobby parliament.

Following a review in late 2006, HM Revenue & Customs has announced changes regarding the administration and collection of excise duty of biofuels and other fuel substitutes (Veg Oil). The changes came into effect on June 30, 2007. There is no longer a requirement to register to pay duty on vegetable oil used as road fuel for those who "produce" or use less than 2,500 litres per year. For those producing over this threshold the biodiesel rate now applies.

Vegetable Oil Economy

Vegetable oil economy deals with the potential of vegetable oil to replace fossil fuels in the economy and how it compares to other potential replacements, including renewable electricity. Vegetable oils are the basis of biodiesel, which can be used like conventional diesel. Some vegetable oil blends are used in unmodified vehicles, but straight vegetable oil needs specially prepared vehicles which have a method of heating the oil to reduce its viscosity and surface tension. Another alternative is vegetable oil refining.

The availability of biodiesel around the world is increasing, although still tiny compared to other fossil fuel sources. There is significant research in algaculture methods to make biofuel from algae. Concerns have been expressed about growing crops for fuel use rather than food and the environmental impacts of large scale agriculture and land clearing required to expand the production of vegetable oil for fuel use. These effects/impacts would need to be specifically researched and evaluated, economically and ecologically, and weighed in balance with the proposed benefits of vegetable oil fuel in relation to the use of other fuel sources.

Limits

The future of vegetable oil as an energy source is limited by the total worldwide oil production. The worldwide production of various types of vegetable oil amounted to 130 million tons in 2008. Its energy content is sufficient to provide 111 million tons of oil equivalent which is 1% of the World energy consumption in 2008 (474 Exajoules or 11320 million tons of oil equivalent).

Future of Energy for World Economy

There is a limited amount of fossil fuel inside the Earth. Since the current world energy resources and consumption is mainly fossil fuels, we are very dependent on them for both transportation and electric power generation. The Hubbert peak theory predicts that oil depletion will result in oil production dropping off in the not too distant future. As time goes on our economy will have to transition to some alternative fuels. Fossil fuels have solved two problems which could be separately solved in the future: the problem of a source of primary energy and of energy storage. Along with straight vegetable oil and biodiesel, some energy technologies that could play an important part in the future future include:

- hydrogen economy
- methanol fuel
- ethanol fuel
- lithium economy
- zinc-air battery
- liquid nitrogen economy
- synthetic fuel
- solar energy / photovoltaics
- nuclear power (fission power)
- fusion power
- wind power
- compressed air energy storage
- flywheel energy storage
- biofuel.

No Net CO_2 or Greenhouse Gas Production

Plants use sunlight and photosynthesis to take carbon dioxide (CO_2) out of the Earth's atmosphere to make vegetable oil. The same CO_2 is then put back after it is burned in an engine. Thus vegetable oil does not increase the CO_2 in the atmosphere, and does not contribute to the problem of greenhouse gas. It is really a way of catching and storing solar energy. It is a true renewable energy.

Burning fossil fuels releases sulphur dioxide (SO_2) and other harmful air pollution. Because vegetable oil has not been inside the earth for millions of years, it is not contaminated with things like sulphur and burns much cleaner, even than ultra-low sulphur diesel. Burning fossil fuels also contributes to the greenhouse gas problem.

Note that if fossil fuels are used in any aspect of production and distribution (making fertilizer, tractors, fuel trucks, etc.), then there would be some contribution to pollution. For it to be 100% non-polluting all aspects of vegetable oil production would have to be non-polluting as well.

Vegetable oil is far less toxic than other fuels such as gasoline, petroleum-based diesel, ethanol, or methanol, and has a much higher flash point (approximately 275-290 °C). The higher flash point reduces the risk of accidental ignition. Some types of vegetable oil are edible.

Generation and Storage

Technologies of hydrogen economy, batteries, compressed air energy storage, and flywheel energy storage address the energy storage problem but not the source of primary energy. Other technologies like fission power, fusion power, and solar power address the problem of a source of primary energy but not energy storage. Vegetable oil addresses both the source of primary energy and of energy storage. The cost and weight to store a given amount of energy as vegetable oil is low compared to many of the potential replacements for fossil fuels.

Type of Vegetable Oil

The list of vegetable oils article discusses which types of vegetable oil are used for fuel and where different types are grown.

Transportation

For transportation the energy density and cost to store the energy are important. If the density is low or the cost is too high it is not practical to make vehicles with reasonable range. Vegetable oil and biodiesel are close to regular diesel.

Another potential issue for new fuels is the Catch-22 conundrum: if there needs to be expensive new infrastructure before people will make cars running on a new fuel, and there needs to be new cars before people will build the infrastructure, how can the transition ever be made? With vegetable oil this is not nearly the problem that it is with some other fuels. The transition from petroleum oil based transportation to vegetable oil based transportation could be gradual and easy compared to hydrogen, ethanol, and most other alternatives. Vegetable oil is used for transportation in four different ways:

- Vegetable oil blends - Mixing vegetable oil with diesel lets users get some of the advantages of burning vegetable oil and is often done with no modification to the vehicle.
- Biodiesel - If vegetable oil is transesterified it becomes biodiesel. Biodiesel burns like normal diesel and works fine in any diesel engine. The name just indicates that the fuel came from vegetable oil.
- Straight vegetable oil - Straight vegetable oil works in diesel engines if it is heated first. Some diesel engines already heat their fuel, others need a small electric heater on the fuel line. How well it works depends on the heating system, the engine, the type of vegetable oil (thinner is easier), and the climate (warmer is easier). Some data is available on results users are

seeing. As vegetable oil has become more popular as a fuel, engines are being designed to handle it better. The Elsbett engine is designed to run on straight vegetable oil. However, as of the start of 2007, it seems that there are not any production vehicles warranties for burning straight vegetable oil, although Deutz offer a tractor and John Deere are known to be in late stages of engine development. There is a German rapeseed oil fuel standard DIN 51605. At this point straight vegetable oil is only a niche market although the market segment in Germany is rapidly growing with large haulage vehicle fleets adopting the fuel, largely for economic reasons. A growing number of decentralised oil mills provide a large part of this fuel.

- Vegetable oil refining - Vegetable oil can be used as feedstock for an oil refinery. There it can be transformed into fuel by hydrocracking (which breaks big molecules into smaller ones using hydrogen) or hydrogenation (which adds hydrogen to molecules). These methods can produce gasoline, diesel, or propane. Some commercial examples of vegetable oil refining are NExBTL, H-Bio, and the ConocoPhilips Process.

The transition can start with biodiesel, vegetable oil refining, and vegetable oil blends, since these technologies do not require the capital outlay of converting an engine to run on vegetable oils. Because it costs to convert vegetable oil into biodiesel it is expected that vegetable oil will always be cheaper than biodiesel. After there are production cars that can use straight vegetable oil and a standard type available at gas stations, consumers will probably choose straight vegetable oil to save money. So the transition to vegetable oil can happen gradually.

Electricity Generation

Vegetable oil is a convenient safe way to store energy for transportation and is similar to the way things have been done. For electricity generation these things are not so important. The most important thing is cost for the electricity produced. The world coal reserves are far larger than the world oil reserves. So replacing the coal used in power plants is not as urgent as replacing the oil used for transportation. The motivation to use vegetable oil for power generation is much less than for transportation. Other methods, like nuclear power, fusion power, wind power and solar power, may provide cheaper electricity, so vegetable oil may only be used in peaking power plants and small power plants, as diesel is limited to today. There is at least one 5 MW power plant that runs on biodiesel.. MAN B&W

Diesel, Wärtsilä and other companies produce engines suitable for power generation that can be fuelled with pure plant oils.

Food vs. Fuel

Food vs. fuel is the dilemma regarding the risk of diverting farmland or crops for biofuels production in detriment of the food supply on a global scale. The "food vs. fuel" or "food or fuel" debate is international in scope, with good and valid arguments on all sides of this issue. There is disagreement about how significant the issue is, what is causing it, and what can or should be done about it.

Biofuel production has increased in recent years. Some commodities like maize (corn), sugar cane or vegetable oil can be used either as food, feed, or to make biofuels. For example, since 2006, a portion of land that was also formerly used to grow other crops in the United States is now used to grow Corn for biofuels, and a larger share of corn is destined to ethanol production, reaching 25% in 2007. A major debate exists on the extent to which biofuels policies contributed to high agricultural prices levels and volatility. A recent study for the International Centre for Trade and Sustainable Development shows that market-driven expansion of ethanol in the US increased maize prices by 21 percent in 2009, in comparison with what prices would have been had ethanol production been frozen at 2004 levels.

Some experts claim that since converting the entire grain harvest of the US would only produce 16% of its auto fuel needs, energy markets are effectively placed in competition with food markets for scarce arable land, resulting in higher food prices. A lot of R&D efforts are currently being put into the production of second generation biofuels from non-food crops, crop residues and waste. Second generation biofuels could hence potentially combine farming for food and fuel and moreover, electricity could be generated simultaneously, which could be beneficial for developing countries and rural areas in developed countries. With global demand for biofuels on the increase due to the oil price increases taking place since 2003 and the desire to reduce oil dependency as well as reduce GHG emissions from transportation, there is also fear of the potential destruction of natural habitats by being converted into farmland. Environmental groups have raised concerns about this trade-off for several years, but now the debate reached a global scale due to the 2007–2008 world food price crisis. On the other hand, several studies do show that biofuel production can be significantly increased without increased acreage. Therefore stating that the crisis in hand relies on the food scarcity.

Brazil has been considered to have the world's first sustainable biofuels economy and its government claims Brazil's sugar cane based ethanol industry has not contributed to the 2008 food crisis. A World Bank policy research working paper released in July 2008 concluded that "...large increases in biofuels production in the United States and Europe are the main reason behind the steep rise in global food prices", and also stated that "Brazil's sugar-based ethanol did not push food prices appreciably higher". However, a 2010 study also by the World Bank concluded that their previous study may have overestimated the contribution of biofuel production, as "the effect of biofuels on food prices has not been as large as originally thought, but that the use of commodities by financial investors (the so-called "financialisation of commodities") may have been partly responsible for the 2007/08 spike." A 2008 independent study by OECD also found that the impact of biofuels on food prices is much smaller.

Food Price Inflation

From 1974 to 2005 real food prices (adjusted for inflation) dropped by 75%. Food commodity prices were relatively stable after reaching lows in 2000 and 2001. Therefore, recent rapid food price increases are considered extraordinary. A World Bank policy research working paper published on July 2008 found that the increase in food commodities prices was led by grains, with sharp price increases in 2005 despite record crops worldwide.

From January 2005 until June 2008, maize prices almost tripled, wheat increased 127 percent, and rice rose 170 percent. The increase in grain prices was followed by increases in fats and oil prices in mid-2006. On the other hand, the study found that sugar cane production has increased rapidly, and it was large enough to keep sugar price increases small except for 2005 and early 2006. The paper concluded that biofuels produced from grains have raised food prices in combination with other related factors between 70 to 75 percent, but ethanol produced from sugar cane has not contributed significantly to the recent increase in food commodities prices.

An economic assessment report published by the OECD in July 2008 found that "...the impact of current biofuel policies on world crop prices, largely through increased demand for cereals and vegetable oils, is significant but should not be overestimated. Current biofuel support measures alone are estimated to increase average wheat prices by about 5 percent, maize by around 7 percent and vegetable oil by about 19 percent over the next 10 years."

Corn is used to make ethanol and prices went up by a factor of three in less than 3 years (measured in US dollars). Reports in 2007 linked stories as diverse as food riots in Mexico due to rising prices of corn for tortillas, and reduced profits at Heineken the large international brewer, to the increasing use of corn (maize) grown in the US Midwest for ethanol production. (In the case of beer, the barley area was cut in order to increase corn production. Barley is not currently used to produce ethanol.) Wheat is up by almost a factor of 3 in 3 years, while soybeans are up by a factor of 2 in 2 years (both measured in US dollars).

As corn is commonly used as feed for livestock, higher corn prices lead to higher prices in animal source foods. Vegetable oil is used to make biodiesel and has about doubled in price in the last couple years. The price is roughly tracking crude oil prices. The 2007–2008 world food price crisis is blamed partly on the increased demand for biofuels. During the same period rice prices went up by a factor of 3 even though rice is not directly used in biofuels.

The USDA expects the 2008/2009 wheat season to be a record crop and 8% higher than the previous year. They also expect rice to have a record crop. Wheat prices have dropped from a high over $12/bushel in May 2008 to under $8/bushel in May. Rice has also dropped from its highs.

According to a 2008 report from the World Bank the production of biofuel pushed food prices up. These conclusions were supported by the Union of Concerned Scientists in their September 2008 newsletter in which they remarked that the World Bank analysis "contradicts U.S. Secretary of Agriculture Ed Schaffer's assertion that biofuels account for only a small percentage of rising food prices."

According to the October Consumer Price Index released Nov. 19, 2008, food prices continued to rise in October 2008 and were 6.3 percent higher than October 2007. Since July 2008 fuel costs dropped by nearly 60 percent.

Proposed Causes

Ethanol Fuel as an Oxygenate Additive

The demand for ethanol fuel produced from field corn was spurred in the U.S. by the discovery that methyl tertiary butyl ether (MTBE) was contaminating groundwater. MTBE use as a oxygenate additive was widespread due to mandates of the Clean Air Act amendments of 1992 to reduce carbon monoxide emissions. As a result, by 2006

MTBE use in gasoline was banned in almost 20 states. There was also concern that widespread and costly litigation might be taken against the U.S. gasoline suppliers, and a 2005 decision refusing legal protection for MTBE, opened a new market for ethanol fuel, the primary substitute for MTBE. At a time when corn prices were around US$ 2 a bushel, corn growers recognised the potential of this new market and delivered accordingly. This demand shift took place at a time when oil prices were already significantly rising.

Government Regulations of Food and Fuel Markets

France, Germany, the United Kingdom and the United States governments have supported biofuels with tax breaks, mandated use, and subsidies. These policies have the unintended consequence of diverting resources from food production and leading to surging food prices and the potential destruction of natural habitats. Fuel for agricultural use often does not have fuel taxes (farmers get duty-free petrol or diesel fuel). Biofuels may have subsidies and low/no retail fuel taxes. Biofuels compete with retail gasoline and diesel prices which have substantial taxes included. The net result is that it is possible for a farmer to use more than a gallon of fuel to make a gallon of biofuel and still make a profit. Some argue that this is a bad distortion of the market. There have been thousands of scholarly papers analysing how much energy goes into making ethanol from corn and how that compares to the energy in the ethanol.

A World Bank policy research working paper concluded that food prices have risen by 35 to 40 percent between 2002–2008, of which 70 to 75 percent is attributable to biofuels. The "month-by-month" five year analysis disputes that increases in global grain consumption and droughts were responsible for significant price increases, reporting that this had had only a marginal impact. Instead the report argues that the EU and US drive for biofuels has had by far the biggest impact on food supply and prices, as increased production of biofuels in the US and EU were supported by subsidies and tariffs on imports, and considers that without these policies, price increases would have been smaller. This research also concluded that Brazil's sugar cane based ethanol has not raised sugar prices significantly, and recommends removing tariffs on ethanol imports by both the US and EU, to allow more efficient producers such as Brazil and other developing countries, including many African countries, to produce ethanol profitably for export to meet the mandates in the EU and the US.

An economic assessment published by the OECD in July 2008 agrees with the World Bank report recommendations regarding the negative effects of subsidies and import tariffs, but found that the estimated impact of biofuels on food prices are much smaller. The OECD study found that trade restrictions, mainly through import tariffs, protect the domestic industry from foreign competitors but impose a cost burden on domestic biofuel users and limits alternative suppliers. The report is also critical of limited reduction of GHG emissions achieved from biofuels based on feedstocks used in Europe and North America, finding that the current biofuel support policies would reduce greenhouse gas emissions from transport fuel by no more than 0.8% by 2015, while Brazilian ethanol from sugar cane reduces greenhouse gas emissions by at least 80% compared to fossil fuels. The assessment calls for the need for more open markets in biofuels and feedstocks in order to improve efficiency and lower costs.

Oil Price Increases

Oil price increases since 2003 resulted in increased demand for biofuels. Transforming vegetable oil into biodiesel is not very hard or costly so there is a profitable arbitrage situation if vegetable oil is much cheaper than diesel. Diesel is also made from crude oil, so vegetable oil prices are partially linked to crude oil prices. Farmers can switch to growing vegetable oil crops if those are more profitable than food crops. So all food prices are linked to vegetable oil prices, and in turn to crude oil prices. A World Bank study concluded that oil prices and a weak dollar explain 25-30% of total price rise between January 2002 until June 2008.

Demand for oil is outstripping the supply of oil and oil depletion is expected to cause crude oil prices to go up over the next 50 years. Record oil prices are inflating food prices worldwide, including those crops that have no relation to biofuels, such as rice and fish.

In Germany and Canada it is now much cheaper to heat a house by burning grain than by using fuel derived from crude oil. With oil at $120/barrel a savings of a factor of 3 on heating costs is possible. When crude oil was at $25/barrel there was no economic incentive to switch to a grain fed heater. From 1971 to 1973, around the time of the 1973 oil crisis, corn and wheat prices went up by a factor of 3. There was no significant biofuel usage at that time.

US Government Policy

Some argue that the US government policy of encouraging ethanol from corn is the main cause for food price increases. US Federal government ethanol subsidies total $7 billion per year, or $1.90 per gallon. Ethanol provides only 55% as much energy as gasoline per gallon, realising about a $3.45 per gallon gasoline trade off. Corn is used to feed chickens, cows, and pigs, so higher corn prices lead to higher prices for chicken, beef, pork, milk, cheese, etc.

U.S. Senators introduced the *BioFuels Security Act* in 2006. "It's time for Congress to realise what farmers in America's heartland have known all along - that we have the capacity and ingenuity to decrease our dependence on foreign oil by growing our own fuel," said U.S. Senator for Illinois Barack Obama.

Two-thirds of U.S. oil consumption is due to the transportation sector. The Energy Independence and Security Act of 2007 has a significant impact on U.S. Energy Policy. With the high profitability of growing corn, more and more farmers switch to growing corn until the profitability of other crops goes up to match that of corn. So the ethanol/corn subsidies drive up the prices of other farm crops.

The US - an important export country for food stocks - will convert 18% of its grain output to ethanol in 2008. Across the US, 25% of the whole corn crop went to ethanol in 2007. The percentage of corn going to biofuel is expected to go up.

Since 2004 a US subsidy has been paid to companies that blend biofuel and regular fuel. The European biofuel subsidy is paid at the point of sale. Companies import biofuel to the US, blend 1% or even 0.1% regular fuel, and then ship the blended fuel to Europe, where it can get a second subsidy. These blends are called B99 or B99.9 fuel. The practice is called "splash and dash". The imported fuel may even come from Europe to the US, get 0.1% regular fuel, and then go back to Europe. For B99.9 fuel the US blender gets a subsidy of $0.999 per gallon. The European biodiesel producers have urged the EU to impose punitive duties on these subsidized imports. US lawmakers are also looking at closing this loophole.

Proposed Action

Freeze on First Generation Biofuel Production

Environmental campaigner George Monbiot has argued for a 5-year freeze on biofuels while their impact on poor communities and the environment is assessed. It has been suggested that a problem with Monbiot's approach is that economic drivers may be required in order to push through the development of more sustainable second-generation biofuel processes: it is possible that these could be stalled if biofuel production decreases. Some environmentalists are suspicious that second-generation biofuels may not solve the problem of a potential clash with food as they also use significant agricultural resources such as water.

A recent UN report on biofuel also raises issues regarding food security and biofuel production. Jean Ziegler, then UN Special Rapporteur on food, concluded that while the argument for biofuels in terms of energy efficiency and climate change are legitimate, the effects for the world's hungry of transforming wheat and maize crops into biofuel are "absolutely catastrophic," and terms such use of arable land a "crime against humanity." Ziegler also calls for a 5-year moratorium on biofuel production. Ziegler's proposal for a five-year ban was rejected by the U.N. Secretary Ban Ki-moon, who called for a comprehensive review of the policies on biofuels, and said that "just criticising biofuel may not be a good solution".

Food surpluses exist in many developed countries. For example, the UK wheat surplus was around 2 million tonnes in 2005. This surplus alone could produce sufficient bioethanol to replace around 2.5% of the UK's petroleum consumption, without requiring any increase in wheat cultivation or reduction in food supply or exports. However, above a few percent, there would be direct competition between first generation biofuel production and food production. This is one reason why many view second generation biofuels as increasingly important.

Non-food Crops for Biofuel

There are different types of biofuels and different feedstocks for them, and it has been proposed that only non-food crops be used for biofuel. This avoids direct competition for commodities like corn and edible vegetable oil. However, as long as farmers can make more money by switching to biofuels they will. The law of supply and demand predicts that if fewer farmers are producing food the price of food will rise.

Second generation biofuels use lignocellulosic raw material such as forest residues (sometimes referred to as brown waste and black liquor from Kraft process or sulfite process pulp mills). Third generation biofuels (biofuel from algae) use non-edible raw materials sources that can be used for biodiesel and bioethanol.

Vegetable Oil Refining

Vegetable oil refining is a process to transform vegetable oil into fuel by hydrocracking or hydrogenation. Hydrocracking breaks big molecules into smaller ones using hydrogen while hydrogenation adds hydrogen to molecules. These methods can be used for production of gasoline, diesel, and propane. Produced diesel fuel is known as *green diesel* or *renewable diesel*. The majority of plant and animal oils are vegetable oils which are triglycerides—suitable for refining. Refinery feedstock includes canola, algae, jatropha, salicornia and tallow. One type of algae, Botryococcus braunii produces a different type of oil, known as a triterpene, which is transformed into alkanes by a different process.

Comparison to Biodiesel

Based on its feedstock green diesel could be classified as biodiesel; however, based on the processing technology and chemical formula green diesel and biodiesel are different products. The chemical reaction commonly-used to produce biodiesel is known as transesterification. Vegetable oil and alcohol are reacted, producing esters, or biodiesel, and the coproduct, glycerol. When refining vegetable oil, no glycerol is produced, only fuels. Refined diesel can be produced that is chemically identical to diesel fuel and does not have the problems specific to transesterified biodiesel. Any blending ratio can be used, and no modifications or checks are required for any diesel engine.

Commercialisation

Some commercial examples of vegetable oil refining are NExBTL, H-Bio, the ConocoPhilips process, and the UOP/Eni Ecofining process. Petrobras planned to use 256 megalitres (1,610,000 bbl) of vegetable oils in the production of H-Bio fuel in 2007. ConocoPhilips is processing 42,000 US gallons per day (1,000 bbl/d) of vegetable oil. Neste Oil completed their first NExBTL plant in the summer 2007 and the second one in 2009. The annual production capacity of each these two plants are 170,000 tons. Two larger refineries with annual capacities of 800,000 tons are built in Rotterdam and Singapore.

Smaller scale refining can be carried out using technology from Renewable Fuel Products, Inc. Germany was expected to replace 3% of their diesel with fuel produced by refining vegetable oil by 2010. In December 2009, the South Dakota School of Mines and Technology received a $1M US grant from the United States government, in part to do research on green diesel, along with biodiesel and other bio fuel research for the United States Air Force. Green diesel was used to power at least one vehicle during a transportation showcase at the 2009 United Nations Climate Change Conference in Copenhagen, Denmark in December 2009.

9

Fragrance Oil

Fragrance oil(s), also known as aroma oils, aromatic oils, and flavour oils, are blended synthetic aroma compounds or natural essential oils that are diluted with a carrier like propylene glycol, vegetable oil, or mineral oil. Aromatic oils are used in perfumery, cosmetics, flavouring of food, and in aromatherapy. To some people, synthetic fragrance oils are less desirable than plant-derived essential oils as components of perfume.

Some include (out of a very diverse range) -

- Ylang ylang
- Vanilla
- Sandalwood
- Cedar
- Mandarin
- Cinnamon
- Lemongrass
- Rosehip
- Peppermint.

Perfume

Perfume is a mixture of fragrant essential oils and/or aroma compounds, fixatives, and solvents used to give the human body, animals, objects, and living spaces "a pleasant scent." The odoriferous compounds that make up a perfume can be manufactured synthetically or extracted from plant or animal sources.

Perfumes have been known to exist in some of the earliest human civilizations, either through ancient texts or from archaeological digs. Modern perfumery began in the late 19th century with the commercial synthesis of aroma compounds such as vanillin or coumarin, which allowed for the composition of perfumes with smells previously unattainable solely from natural aromatics alone.

The word perfume used today derives from the Latin *per fumus*, meaning "through smoke." Perfumery, or the art of making perfumes, began in ancient Mesopotamia and Egypt and was further refined by the Romans and Persians. Although perfume and perfumery also existed in India, much of its fragrances are incense based. The earliest distillation of Ittar, Arabic meaning scent, was mentioned in the Hindu Ayurvedic text Charaka Samhita. The Harshacharita, written in 7th century in Northern India mentions use of fragrant agarwood oil.

The world's first recorded chemist is considered to be a woman named Tapputi, a perfume maker who was mentioned in a cuneiform tablet from the 2nd millennium BC in Mesopotamia. She distilled flowers, oil, and calamus with other aromatics then filtered and put them back in the still several times.

In 2005, archaeologists uncovered what are believed to be the world's oldest perfumes in Pyrgos, Cyprus. The perfumes date back more than 4,000 years. The perfumes were discovered in an ancient perfumery. At least 60 stills, mixing bowls, funnels and perfume bottles were found in the 43,000-square-foot (4,000 m^2) factory. In ancient times people used herbs and spices, like almond, coriander, myrtle, conifer resin, bergamot, as well as flowers.

The Arabian chemist, Al-Kindi (Alkindus), wrote in the 9th century a book on perfumes which he named *Book of the Chemistry of Perfume and Distillations*. It contained more than a hundred recipes for fragrant oils, salves, aromatic waters and substitutes or imitations of costly drugs. The book also described 107 methods and recipes for perfume-making and perfume making equipment, such as the alembic (which still bears its Arabic name).

The Persian chemist Ibn Sina (also known as Avicenna) introduced the process of extracting oils from flowers by means of distillation, the procedure most commonly used today. He first experimented with the rose. Until his discovery, liquid perfumes were mixtures of oil and crushed herbs or petals, which made a strong blend. Rose water was more delicate, and immediately became popular. Both of the raw ingredients and distillation technology significantly influenced western perfumery and scientific developments, particularly chemistry. The art of perfumery was known in western Europe ever since 1221, if we consider the monks' recipes of Santa Maria delle Vigne or Santa Maria Novella of Florence, Italy.

In the east, the Hungarians produced in 1370 a perfume made of scented oils blended in an alcohol solution at the command of Queen Elizabeth of Hungary, best known as Hungary Water. The art of perfumery prospered in Renaissance Italy, and in the 16th century, Italian refinements were taken to France by Catherine de' Medici's personal perfumer, Rene the Florentine (Renato il fiorentino). His laboratory was connected with her apartments by a secret passageway, so that no formulas could be stolen en route. Thanks to Rene, France quickly became one of the European centres of perfume and cosmetic manufacture. Cultivation of flowers for their perfume essence, which had begun in the 14th century, grew into a major industry in the south of France. Between the 16th and 17th century, perfumes were used primarily by the wealthy to mask body odours resulting from infrequent bathing. Partly due to this patronage, the perfumery industry was created. In Germany, Italian barber Giovanni Paolo Feminis created a perfume water called Aqua Admirabilis, today best known as eau de cologne, while his nephew Johann Maria Farina (Giovanni Maria Farina) in 1732 took over the business. By the 18th century, aromatic plants were being grown in the Grasse region of France, in Sicily, and in Calabria, Italy to provide the growing perfume industry with raw materials. Even today, Italy and France remain the centre of the European perfume design and trade.

Concentration

Perfume types reflect the concentration of aromatic compounds in a solvent, which in fine fragrance is typically ethanol or a mix of water and ethanol. Various sources differ considerably in the definitions of perfume types. The intensity and longevity of a perfume is based on the concentration, intensity and longevity of the aromatic compounds (natural essential oils / perfume oils) used: As the percentage of aromatic compounds increases, so does the intensity and longevity of the scent created. Specific terms are used to describe a fragrance's approximate concentration by percent/volume of perfume oil, which are typically vague or imprecise. A list of common terms (Perfume-Classification) is as follows:

- Perfume extract, or simply perfume (Extrait): 15-40% (IFRA: typical 20%) aromatic compounds

- Esprit de Parfum (ESdP): 15-30% aromatic compounds, a seldom used strength concentration in between EdP and perfume
- Eau de Parfum (EdP), Parfum de Toilette (PdT): 10-20% (typical ~15%) aromatic compounds, sometimes listed as "eau de perfume" or "millésime." Parfum de Toilette is a less common term that is generally analogous to Eau de Parfum.
- Eau de toilette (EdT): 5-15% (typical ~10%) aromatic compounds
- Eau de Cologne (EdC): Chypre citrus type perfumes with 3-8% (typical ~5%) aromatic compounds. "Original Eau de Cologne" is a registered trademark.
- Perfume mist: 3-8% aromatic compounds (typical non-alcohol solvent)
- Splash (EdS) and Aftershave: 1-3% aromatic compounds. "EdS" is a registered trademark.

Solvent Types

Perfume oils are often diluted with a solvent, though this is not always the case, and its necessity is disputed. By far the most common solvent for perfume oil dilution is ethanol or a mixture of ethanol and water. Perfume oil can also be diluted by means of neutral-smelling oils such as fractionated coconut oil, or liquid waxes such as jojoba oil.

Imprecise Terminology

Although quite often *Eau de Parfum* (EdP) will be more concentrated than *Eau de Toilette* (EdT) and in turn *Eau de Cologne* (EdC), this is not always the case. Different perfumeries or perfume houses assign different amounts of oils to each of their perfumes. Therefore, although the oil concentration of a perfume in EdP dilution will necessarily be higher than the same perfume in EdT from within the same range, the actual amounts can vary between perfume houses. An EdT from one house may be stronger than an EdP from another.

Men's fragrances are rarely sold as EdP or perfume extracts; equally so, women's fragrances are rarely sold in EdC concentrations. Although this gender specific naming trend is common for assigning fragrance concentrations, it does not directly have anything to do with whether a fragrance was intended for men or women. Furthermore, some fragrances with the same *product name* but having a different *concentration name* may not only differ in their dilutions, but actually use different perfume oil mixtures altogether. For instance, in order to make the EdT version of a fragrance brighter and fresher than its

EdP, the EdT oil may be "tweaked" to contain slightly more top notes or fewer base notes. In some cases, words such as *extrême*, *intense*, or *concentrée* that might indicate aromatic concentration are actually completely different fragrances, related only because of a similar perfume *accord*. An example of this is Chanel's *Pour Monsieur* and *Pour Monsieur Concentrée*.

Eau de Cologne (EdC) since 1706 in Cologne, Germany, is originally a specific fragrance and trademark. However outside of Germany the term has become generic for Chypre citrus perfumes (without base-notes). EdS (since 1993) is a new perfume class and a registered trademark.

Describing a Perfume

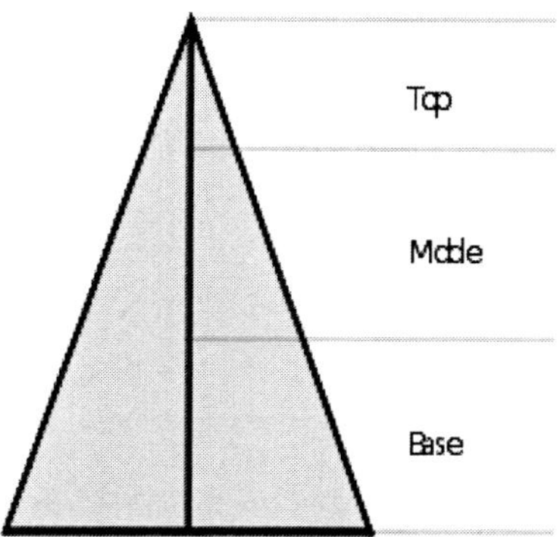

***Figure** : Fragrance pyramid*

The precise formulae of commercial perfumes are kept secret. Even if they were widely published, they would be dominated by such complex ingredients and odourants that they would be of little use in providing a guide to the general consumer in description of the *experience* of a scent. Nonetheless, connoisseurs of perfume can become extremely skillful at identifying components and origins of scents in the same manner as wine experts.

The most practical way to start describing a perfume is according to the elements of the *fragrance notes* of the scent or the "family" it belongs to, all of which affect the overall impression of a perfume from first application to the last lingering hint of scent.

Fragrance Notes

Perfume is described in a musical metaphor as having three sets of *notes*, making the harmonious scent *accord*. The notes unfold over time, with the immediate impression of the top note leading to the deeper middle notes, and the base notes gradually appearing as the final stage. These notes are created carefully with knowledge of the evapouration process of the perfume.

- Top notes: The scents that are perceived immediately on application of a perfume. Top notes consist of small, light molecules that evapourate quickly. They form a person's initial impression of a perfume and thus are very important in the selling of a perfume. Also called the *head notes.*
- Middle notes: The scent of a perfume that emerges just prior to when the top notes dissipate. The middle note compounds form the "heart" or main body of a perfume and act to mask the often unpleasant initial impression of base notes, which become more pleasant with time. They are also called the *heart notes.*
- Base notes: The scent of a perfume that appears close to the departure of the middle notes. The base and middle notes together are the main theme of a perfume. Base notes bring depth and solidity to a perfume. Compounds of this class of scents are typically rich and "deep" and are usually not perceived until 30 minutes after application.

The scents in the top and middle notes are influenced by the base notes, as well the scents of the base notes will be altered by the type of fragrance materials used as middle notes. Manufacturers of perfumes usually publish perfume notes and typically they present it as fragrance pyramid, with the components listed in imaginative and abstract terms.

Olfactive Families

Grouping perfumes, like any taxonomy, can never be a completely objective or final process. Many fragrances contain aspects of different families. Even a perfume designated as "single flower", however subtle, will have undertones of other aromatics. "True" unitary scents can rarely be found in perfumes as it requires the perfume to exist only as a singular aromatic material. Classification by olfactive family is a starting point for a description of a perfume, but it cannot by itself denote the specific characteristic of that perfume.

Traditional

The traditional classification which emerged around 1900 comprised the following categories:

- Single Floral: Fragrances that are dominated by a scent from one particular flower; in French called a *soliflore.* (e.g. Serge Lutens' *Sa Majeste La Rose*, which is dominated by rose.)

- Floral Bouquet: Is a combination of fragrance of several flowers in a perfume compound. Examples include *Quelques Fleurs* by Houbigant and *Joy* by Jean Patou.
- Ambered, or "Oriental": A large fragrance class featuring the sweet slightly animalic scents of ambergris or labdanum, often combined with vanilla, tonka bean, flowers and woods. Can be enhanced by camphorous oils and incense resins, which bring to mind Victorian era imagery of the Middle East and Far East. Traditional examples include Guerlain's *Shalimar* and Yves Saint Laurent's *Opium*.
- Wood: Fragrances that are dominated by woody scents, typically of agarwood, sandalwood and cedarwood. Patchouli, with its camphoraceous smell, is commonly found in these perfumes. A traditional example here would be Myrurgia's *Maderas De Oriente* or Chanel *Bois-des-Îles*. A modern example would be Balenciaga *Rumba*.
- Leather: A family of fragrances which features the scents of honey, tobacco, wood and wood tars in its middle or base notes and a scent that alludes to leather. Traditional examples include Robert Piguet's *Bandit* and Balmain's *Jolie Madame*.
- Chypre: Meaning *Cyprus* in French, this includes fragrances built on a similar accord consisting of bergamot, oakmoss, patchouli, and labdanum. This family of fragrances is named after a perfume by François Coty, and one of the most famous examples is Guerlain's *Mitsouko*.
- Fougère: Meaning *Fern* in French, built on a base of lavender, coumarin and oakmoss. Houbigant's *Fougère Royale* pioneered the use of this base. Many men's fragrances belong to this family of fragrances, which is characterised by its sharp herbaceous and woody scent. Some well-known modern fougères are Fabergé *Brut* and Guy Laroche *Drakkar Noir*.

Modern

Since 1945, due to great advances in the technology of perfume creation as well as the natural development of styles and tastes; new categories have emerged to describe modern scents:

- Bright Floral: combining the traditional Single Floral & Floral Bouquet categories. A good example would be Estée Lauder's *Beautiful*.
- Green: a lighter and more modern interpretation of the Chypre type, with pronounced cut grass, crushed green leaf and cucumber-like scents. Two examples would be Estée Lauder's *Aliage* or Sisley's *Eau de Campagne*.

- Aquatic, Oceanic, or Ozonic: the newest category in perfume history, appearing in 1991 with Christian Dior's *Dune*. A very clean, modern smell leading to many of the modern androgynous perfumes. Generally contains calone, a synthetic scent discovered in 1966. Also used to accent floral, oriental, and woody fragrances.
- Citrus: An old fragrance family that until recently consisted mainly of "freshening" eau de colognes, due to the low tenacity of citrus scents. Development of newer fragrance compounds has allowed for the creation of primarily citrus fragrances. A good example here would be *Brut*.
- Fruity: featuring the aromas of fruits other than citrus, such as peach, cassis (black currant), mango, passion fruit, and others. A modern example here would be Ginestet *Botrytis*.
- Gourmand (French: [au□mQ]): scents with "edible" or "dessert"-like qualities. These often contain notes like vanilla, tonka bean and coumarin, as well as synthetic components designed to resemble food flavours. A sweet example is Thierry Mugler's *Angel*. A savory example would be *Dinner* by BoBo, which has cumin and curry hints.

Fragrance Wheel

The Fragrance wheel is a relatively new classification method that is widely used in retail and in the fragrance industry. The method was created in 1983 by Michael Edwards, a consultant in the perfume industry, who designed his own scheme of fragrance classification. The new scheme was created in order to simplify fragrance classification and naming scheme, as well as to show the relationships between each of the individual classes.

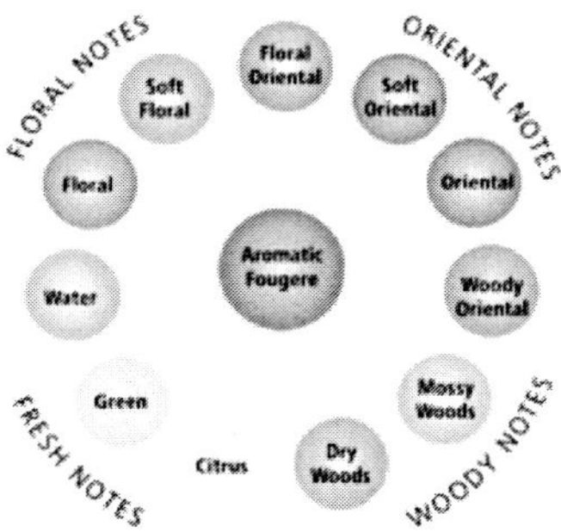

***Figure** : Fragrance Wheel perfume classification chart, ver. 1983*

The five standard families consist of *Floral*, *Oriental*, *Woody*, *Fougère*, and *Fresh*, with the former four families being more "classic" while the latter consisting of newer bright and clean smelling citrus and oceanic fragrances that have arrived due to improvements in fragrance technology. Each of the families are in turn divided into sub-groups and arranged around a wheel.

Aromatics Sources

Plant Sources

Plants have long been used in perfumery as a source of essential oils and aroma compounds. These aromatics are usually secondary metabolites produced by plants as protection against herbivores, infections, as well as to attract pollinators.

Plants are by far the largest source of fragrant compounds used in perfumery. The sources of these compounds may be derived from various parts of a plant. A plant can offer more than one source of aromatics, for instance the aerial portions and seeds of coriander have remarkably different odours from each other. Orange leaves, blossoms, and fruit zest are the respective sources of petitgrain, neroli, and orange oils.

- Bark: Commonly used barks includes cinnamon and cascarilla. The fragrant oil in sassafras root bark is also used either directly or purified for its main constituent, safrole, which is used in the synthesis of other fragrant compounds.
- Flowers and blossoms: Undoubtedly the largest and most common source of perfume aromatics. Includes the flowers of several species of rose and jasmine, as well as osmanthus, plumeria, mimosa, tuberose, narcissus, scented geranium, cassie, ambrette as well as the blossoms of citrus and ylang-ylang trees. Although not traditionally thought of as a flower, the unopened flower buds of the clove are also commonly used. Most orchid flowers are not commercially used to produce essential oils or absolutes, except in the case of vanilla, an orchid, which must be pollinated first and made into seed pods before use in perfumery.
- Fruits: Fresh fruits such as apples, strawberries, cherries unfortunately do not yield the expected odours when extracted; if such fragrance notes are found in a perfume, they are synthetic. Notable exceptions include litsea cubeba, vanilla, and juniper berry. The most commonly used fruits yield their

aromatics from the rind; they include citrus such as oranges, lemons, and limes. Although grapefruit rind is still used for aromatics, more and more commercially used grapefruit aromatics are artificially synthesized since the natural aromatic contains sulphur and its degradation product is quite unpleasant in smell.

- Leaves and twigs: Commonly used for perfumery are lavender leaf, patchouli, sage, violets, rosemary, and citrus leaves. Sometimes leaves are valued for the "green" smell they bring to perfumes, examples of this include hay and tomato leaf.
- Resins: Valued since antiquity, resins have been widely used in incense and perfumery. Highly fragrant and antiseptic resins and resin-containing perfumes have been used by many cultures as medicines for a large variety of ailments. Commonly used resins in perfumery include labdanum, frankincense/olibanum, myrrh, Peru balsam, gum benzoin. Pine and fir resins are a particularly valued source of terpenes used in the organic synthesis of many other synthetic or naturally occurring aromatic compounds. Some of what is called amber and copal in perfumery today is the resinous secretion of fossil conifers.
- Roots, rhizomes and bulbs: Commonly used terrestrial portions in perfumery include iris rhizomes, vetiver roots, various rhizomes of the ginger family.
- Seeds: Commonly used seeds include tonka bean, carrot seed, coriander, caraway, cocoa, nutmeg, mace, cardamom, and anise.
- Woods: Highly important in providing the base notes to a perfume, wood oils and distillates are indispensable in perfumery. Commonly used woods include sandalwood, rosewood, agarwood, birch, cedar, juniper, and pine. These are used in the form of macerations or dry-distilled (rectified) forms.

Ambergris

- Ambergris: Lumps of oxidised fatty compounds, whose precursors were secreted and expelled by the sperm whale. Ambergris should not be confused with yellow amber, which is used in jewellery. Because the harvesting of ambergris involves no harm to its animal source, it remains one of the few animalic fragrancing agents around which little controversy now exists.
- Castoreum: Obtained from the odorous sacs of the North American beaver.

- Civet: Also called Civet Musk, this is obtained from the odorous sacs of the civets, animals in the family *Viverridae*, related to the mongoose. The World Society for the Protection of Animals investigated African civets caught for this purpose.
- Hyraceum: Commonly known as "Africa Stone", is the petrified excrement of the Rock Hyrax.
- Honeycomb: From the honeycomb of the honeybee. Both beeswax and honey can be solvent extracted to produce an absolute. Beeswax is extracted with ethanol and the ethanol evapourated to produce beeswax absolute.
- Deer musk: Originally derived from the musk sacs from the Asian musk deer, it has now been replaced by the use of synthetic musks sometimes known as "white musk".

Other natural sources:

- Lichens: Commonly used lichens include oakmoss and treemoss thalli.
- "Seaweed": Distillates are sometimes used as essential oil in perfumes. An example of a commonly used seaweed is *Fucus vesiculosus*, which is commonly referred to as bladder wrack. Natural seaweed fragrances are rarely used due to their higher cost and lower potency than synthetics.

Synthetic Sources

Many modern perfumes contain synthesized odourants. Synthetics can provide fragrances which are not found in nature. For instance, Calone, a compound of synthetic origin, imparts a fresh ozonous metallic marine scent that is widely used in contemporary perfumes. Synthetic aromatics are often used as an alternate source of compounds that are not easily obtained from natural sources. For example, linalool and coumarin are both naturally occurring compounds that can be inexpensively synthesized from terpenes. Orchid scents (typically *salicylates*) are usually not obtained directly from the plant itself but are instead synthetically created to match the fragrant compounds found in various orchids.

One of the most commonly used class of synthetic aromatic by far are the white musks. These materials are found in all forms of commercial perfumes as a neutral background to the middle notes. These musks are added in large quantities to laundry detergents in order to give washed clothes a lasting "clean" scent.

The majority of the world's synthetic aromatics are created by relatively few companies. They include:

- International Flavours and Fragrances (IFF)
- Givaudan
- Firmenich
- Takasago
- Symrise.

Each of these companies patents several processes for the production of aromatic synthetics annually.

Obtaining Natural Odourants

Before perfumes can be composed, the odourants used in various perfume compositions must first be obtained. Synthetic odourants are produced through organic synthesis and purified. Odourants from natural sources require the use of various methods to extract the aromatics from the raw materials. The results of the extraction are either essential oils, absolutes, concretes, or butters, depending on the amount of waxes in the extracted product.

All these techniques will, to a certain extent, distort the odor of the aromatic compounds obtained from the raw materials. This is due to the use of heat, harsh solvents, or through exposure to oxygen in the extraction process which will denature the aromatic compounds, which either change their odor character or renders them odourless.

- Maceration/Solvent extraction: The most used and economically important technique for extracting aromatics in the modern perfume industry. Raw materials are submerged in a solvent that can dissolve the desired aromatic compounds. *Maceration* lasts anywhere from hours to months. Fragrant compounds for woody and fibrous plant materials are often obtained in this manner as are all aromatics from animal sources. The technique can also be used to extract odourants that are too volatile for *distillation* or easily denatured by heat. Commonly used solvents for *maceration/solvent extraction* include hexane, and dimethyl ether. The product of this process is called a "concrete."
    - o *Supercritical fluid extraction*: A relatively new technique for extracting fragrant compounds from a raw material, which often employs Supercritical CO_2. Due to the low heat of process and the relatively nonreactive solvent used in the extraction, the fragrant compounds derived often closely resemble the original odor of the raw material.
    - o *Ethanol extraction*: A type of solvent extraction used to extract fragrant compounds directly from dry raw materials,

as well as the impure oily compounds materials resulting from solvent extraction or enfleurage. Ethanol extraction is not used to extract fragrance from fresh plant materials since these contain large quantities of water, which will also be extracted into the ethanol.

- Distillation: A common technique for obtaining aromatic compounds from plants, such as orange blossoms and roses. The raw material is heated and the fragrant compounds are re-collected through condensation of the distilled vapour.
  - o *Steam distillation*: Steam from boiling water is passed through the raw material, which drives out their volatile fragrant compounds. The condensate from distillation is settled in a Florentine flask. This allows for the easy separation of the fragrant oils from the water. The water collected from the condensate, which retains some of the fragrant compounds and oils from the raw material is called hydrosol and sometimes sold. This is most commonly used for fresh plant materials such as flowers, leaves, and stems.
  - o *Dry/destructive distillation*: The raw materials are directly heated in a still without a carrier solvent such as water. Fragrant compounds that are released from the raw material by the high heat often undergo anhydrous pyrolysis, which results in the formation of different fragrant compounds, and thus different fragrant notes. This method is used to obtain fragrant compounds from fossil amber and fragrant woods where an intentional "burned" or "toasted" odor is desired.
  - o *Fractionation*: Through the use of a fractionation column, different fractions distilled from a material can be selectively excluded to modify the scent of the final product. Although the product is more expensive, this is sometimes performed to remove unpleasant or undesirable scents of a material and affords the perfumer more control over their composition process.
- Expression: Raw material is squeezed or compressed and the oils are collected. Of all raw materials, only the fragrant oils from the peels of fruits in the citrus family are extracted in this manner since the oil is present in large enough quantities as to make this extraction method economically feasible.

- Enfleurage: Absorption of aroma materials into solid fat or wax and then extracting the odorous oil with ethyl alcohol. Extraction by enfleurage was commonly used when distillation was not possible because some fragrant compounds denature through high heat. This technique is not commonly used in the present day industry due to its prohibitive cost and the existence of more efficient and effective extraction methods.

Composing Perfumes

Perfume compositions are an important part of many industries ranging from the luxury goods sectors, food services industries, to manufacturers of various household chemicals. The purpose of using perfume or fragrance compositions in these industries is to affect customers through their sense of smell and entice them into purchasing the perfume or perfumed product. As such there is significant interest in producing a perfume formulation that people will find aesthetically pleasing.

The Perfumer

The job of composing perfumes that will be sold is left up to an expert on perfume composition or known in the fragrance industry as the *perfumer*. They are also sometimes referred to affectionately as a "*Nez*" (French for *nose*) due to their fine sense of smell and skill in smell composition.

The composition of a perfume typically begins with a *brief* by the perfumer's employer or an outside customer. The customers to the perfumer or their employers, are typically fashion houses or large corporations of various industries. The perfumer will then go through the process of blending multiple perfume mixtures and sell the formulation to the customer, often with modifications of the composition of the perfume. The perfume composition will then be either used to enhance another product as a *functional fragrance* (shampoos, make-up, detergents, car interiors, etc.), or marketed and sold directly to the public as a *fine fragrance*.

Technique

Although there is no single "correct" technique for the formulation of a perfume, there are general guidelines as to how a perfume can be constructed from a concept. Although many ingredients do not contribute to the smell of a perfume, many perfumes include colourants and anti-oxidants to improve the marketability and shelf life of the perfume, respectively.

Basic Framework

Perfume oils usually contain tens to hundreds of ingredients and these are typically organised in a perfume for the specific role they will play. These ingredients can be roughly grouped into four groups:

- *Primary scents* (Heart): Can consist of one or a few main ingredients for a certain concept, such as "rose". Alternatively, multiple ingredients can be used together to create an "abstract" primary scent that does not bear a resemblance to a natural ingredient. For instance, jasmine and rose scents are commonly blends for abstract floral fragrances. Cola flavourant is a good example of an abstract primary scent.
- *Modifiers*: These ingredients alter the primary scent to give the perfume a certain desired character: for instance, fruit esters may be included in a floral primary to create a fruity floral; calone and citrus scents can be added to create a "fresher" floral. The cherry scent in cherry cola can be considered a modifier.
- *Blenders*: A large group of ingredients that smooth out the transitions of a perfume between different "layers" or bases. These themselves can be used as a major component of the primary scent. Common blending ingredients include linalool and hydroxycitronellal.
- *Fixatives*: Used to support the primary scent by bolstering it. Many resins, wood scents, and amber bases are used as fixatives.

The top, middle, and base notes of a fragrance may have separate primary scents and supporting ingredients. The perfume's fragrance oils are then blended with ethyl alcohol and water, aged in tanks for several weeks and filtered through processing equipment to, respectively allow the perfume ingredients in the mixture to stabilise and to remove any sediment and particles before the solution can be filled into the perfume bottles.

Fragrance Bases

Instead of building a perfume from "ground up", many modern perfumes and colognes are made using *fragrance bases* or simply bases. Each base is essentially modular perfume that is blended from essential oils and aromatic chemicals, and formulated with a simple concept such as "fresh cut grass" or "juicy sour apple". Many of Guerlain's *Aqua Allegoria* line, with their simple fragrance concepts, are good examples of what perfume fragrance bases are like.

The effort used in developing bases by fragrance companies or individual perfumers may equal that of a marketed perfume, since they are useful in that they are reusable. On top of its reusability, the benefit in using bases for construction is quite numerous:

1. Ingredients with "difficult" or "overpowering" scents that are tailored into a blended base may be more easily incorporated into a work of perfume
2. A base may be better scent approximations of a certain thing than the extract of the thing itself. For example, a base made to embody the scent for "fresh dewy rose" might be a better approximation for the scent concept of a rose after rain than plain rose oil. Flowers whose scents cannot be extracted, such as gardenia or hyacinth, are composed as bases from data derived from headspace technology.
3. A perfumer can quickly rough out a concept from a brief by cobbling together multiple bases, then present it feedback. Smoothing out the "edges" of the perfume can be done after a positive response.

Reverse Engineering

Creating perfumes through reverse engineering with analytical techniques such as GC/MS can reveal the "general" formula for any particular perfume. The difficulty of GC/MS analysis arises due to the complexity of a perfume's ingredients. This is particularly due to the presence of natural essential oils and other ingredients consisting of complex chemical mixtures. However, "anyone armed with good GC/MS equipment and experienced in using this equipment can today, within days, find out a great deal about the formulation of any perfume... customers and competitors can analyse most perfumes more or less precisely."

Antique or badly preserved perfumes undergoing this analysis can also be difficult due to the numerous degradation by-products and impurities that may have resulted from breakdown of the odorous compounds. Ingredients and compounds can usually be ruled out or identified using gas chromatograph (GC) smellers, which allow individual chemical components to be identified both through their physical properties and their scent. Reverse engineering of best-selling perfumes in the market is a very common practice in the fragrance industry due to the relative simplicity of operating GC equipment, the pressure to produce marketable fragrances, and the highly lucrative nature of the perfume market.

Health and Environmental Issues

Perfume ingredients, regardless of natural or synthetic origins, may all cause health or environmental problems when used or abused in substantial quantities. Although the areas are under active research, much remains to be learned about the effects of fragrance on human health and the environment.

Immunological

Evidence in peer-reviewed journals shows that some fragrances can cause asthmatic reactions in some individuals, especially those with severe or atopic asthma. Many fragrance ingredients can also cause headaches, allergic skin reactions or nausea.

In some cases, an excessive use of perfumes may cause allergic reactions of the skin. For instance, acetophenone, ethyl acetate and acetone while present in many perfumes, are also known or potential respiratory allergens. Nevertheless this may be misleading, since the harm presented by many of these chemicals (either natural or synthetic) is dependent on environmental conditions and their concentrations in a perfume. For instance, linalool, which is listed as an irritant, causes skin irritation when it degrades to peroxides, however the use of antioxidants in perfumes or reduction in concentrations can prevent this. As well, the furanocoumarin present in natural extracts of grapefruit or celery can cause severe allergic reactions and increase sensitivity to ultraviolet radiation. Some research on natural aromatics has shown that many contain compounds that cause skin irritation. However some studies, such as IFRA's research claim that opoponax is too dangerous to be used in perfumery, still lack scientific consensus. It is also true that sometimes inhalation alone can cause skin irritation.

Carcinogenicity

There is scientific evidence that nitro-musks such as Musk xylene can cause cancer while common ingredients, like certain polycyclic synthetic musks, can disrupt the balance of hormones in the human body (endocrine disruption). Some natural aromatics, such as oakmoss absolutes, contain allergens and carcinogenic compounds.

Toxicity

Certain chemicals found in perfume are often toxic, at least for small insects if not for humans. For example the compound Tricyclodecenyl allyl ether is often found in synthetic perfumesand has insect repellent property.

Species Endangerment

The demands for aromatic materials like sandalwood, agarwood, musk has led to the endangerment of these species as well as illegal trafficking and harvesting.

Safety Regulation

The perfume industry in the US is not directly regulated by the FDA, instead the FDA controls the safety of perfumes through their ingredients and requires that they be tested to the extent that they are *Generally recognised as safe* (GRAS).

Due to the need for protection of trade secrets, companies rarely give the full listing of ingredients regardless of their effects on health. In Europe, as from 11 March 2005, the mandatory listing of a set of 26 recognised fragrance allergens was enforced. The requirement to list these materials is dependant on the intended use of the final product. The limits above which the allegens are required to be declared are 0.001% for products intended to remain on the skin, and 0.01% for those intended to be rinsed off. This has resulted in many old perfumes like chypres and fougère classes, which require the use of oakmoss extract, being reformulated.

Preserving Perfume

Fragrance compounds in perfumes will degrade or break down if improperly stored in the presence of:

- Heat
- Light
- Oxygen
- Extraneous organic materials.

Proper preservation of perfumes involve keeping them away from sources of heat and storing them where they will not be exposed to light. An opened bottle will keep its aroma intact for several years, as long as it is well stored. However the presence of oxygen in the head space of the bottle and environmental factors will in the long run alter the smell of the fragrance.

Perfumes are best preserved when kept in light-tight aluminium bottles or in their original packaging when not in use, and refrigerated to relatively low temperatures: between 3-7 degrees Celsius (37-45 degrees Fahrenheit). The Osmothèque, a perfume conservatory and museum, store their perfumes in argon evacuated aluminium flasks at 12 degrees Celsius. Although it is difficult to completely remove oxygen from the headspace of a stored flask of fragrance, opting for spray dispensers instead of rollers and "open" bottles will minimise oxygen exposure. Sprays also have the advantage of isolating fragrance inside a bottle and preventing it from mixing with dust, skin, and detritus, which would degrade and alter the quality of a perfume.

Perfumer

A perfumer is a term used for an expert on creating perfume compositions, sometimes referred to affectionately as a *Nose* (French: *le nez*) due to their fine sense of smell and skill in producing olfactory compositions. The perfumer is effectively an artist who is trained in depth on the concepts of fragrance aesthetics and who is capable of conveying abstract concepts and moods with fragrance compositions. At the most rudimentary level, a perfumer must have a keen knowledge of a large variety of fragrance ingredients and their smells, and be able to distinguish each of the fragrance ingredients whether alone or in combination with other fragrances. As well, they must know how each ingredient reveals itself through time with other ingredients. The job of the perfumer is very similar to that of flavourists, who compose smells and flavourants for many commercial food products.

Employment

Most perfumers are employed by several large fragrance corporations in the world including Mane, Firmenich, IFF, Givaudan, Takasago, and Symrise. Some perfumers work exclusively for a perfume house or in their own company, but these cases are not as common.

The perfumer typically begins a perfume project with a *brief* by the perfumer's employer or an outside customer. The customers to the perfumer or their employers, are typically fashion houses or large corporations of various industries. Each brief will contain the specifications for the desired perfume, and will describe in often poetic or abstract terms what the perfume should smell like or what feelings it should evoke in those who smell it, along with a maximum per litre price of the perfume oil concentrate. This allowance, along with the intended application of the perfume, will determine what aromatic ingredients will be used in the perfume composition.

The perfumer will then go through the process of blending multiple perfume mixtures and will attempt to capture the desired feelings specified in the brief. After presenting the perfume mixtures to the customers, the perfumer may "win" the brief with their approval. They proceed to work with the customer, often with the direction provided by a panel or artistic director, which guides and edits the modifications on the composition of the perfume. This process typically spans several months to several years, going over many iterations and may involve cultural and public surveys to tailor a perfume to a particular market. The perfume composition will then be either used to enhance another product as a *functional fragrance* (shampoos, make-up, detergents, car interiors, etc.) or marketed and sold directly to the public as a *fine fragrance.* Alternatively, the perfumer may simply be inspired to create a perfume and produce something that later becomes marketable or wins a brief. This is more common in smaller or independent perfume houses.

History of Perfume

The history of perfume began in antiquity. The word *perfume* is used today to describe scented mixtures and is derived from the Latin word, "*per fumus*", meaning *through smoke.* Perfumery, or the art of making perfumes, began in ancient Egypt but was developed and further refined by the Romans and the Arabs. Although perfume and perfumery also existed in East Asia, much of its fragrances are incense based. The basic ingredients and methods of making perfumes are described by Pliny the Elder in his Naturalis Historia.

The world's first recorded chemist is a person named Tapputi, a perfume maker who was mentioned in a Cuneiform tablet from the 2nd millennium BC in Mesopotamia. Perfume is thousands of years old and is made from fragrance oils. Perfume and perfumery also existed in India, much of its fragrances were incense based. The earliest distillation of Attar was mentioned in the Hindu Ayurvedic text Charaka Samhita. The Harshacharita, written in 7th century A.D. in Northern India mentions use of fragrant agarwood oil.

Aroma Compound

An aroma compound, also known as odourant, aroma, fragrance or flavour, is a chemical compound that has a smell or odor. A chemical compound has a smell or odor when two conditions are met: the compound needs to be volatile, so it can be transported to the olfactory system in the upper part of the nose, and it needs to be in a sufficiently high concentration to be able to interact with one or more of the olfactory receptors.

Aroma compounds can be found in food, wine, spices, perfumes, fragrance oils, and essential oils. For example, many form biochemically during ripening of fruits and other crops. In wines, most form as byproducts of fermentation. Odourants can also be added to a dangerous odourless substance, like propane, natural gas, or hydrogen, as a warning. Also, many of the aroma compounds play a significant role in the production of flavourants, which are used in the food service industry to flavour, improve, and generally increase the appeal of their products.

Cooking Oil

Cooking oil is fat that is used for cooking and is usually liquid at room temperature. Some saturated oils such as coconut oil and palm oil are more solid at room temperature than others. Cooking oil may be of plant or animal origin. Types of cooking oil include: ghee, olive oil, palm oil, soybean oil, canola oil, pumpkin seed oil, corn oil, sunflower oil, safflower oil, peanut oil, grape seed oil, sesame oil, argan oil, rice bran oil and other vegetable oils. Oil can be flavoured with aromatic foodstuffs such as herbs, chillies or garlic.

Health and Nutrition

The appropriate amount of fat as a component of daily food consumption is the topic of some controversy. Some fat is required in the diet, and fat (in the form of oil) is also essential in many types of cooking. The FDA recommends that 30% or less of calories consumed daily should be from fat. Other nutritionists recommend that no more than 10% of a person's daily calories come from fat. In extremely cold environments, a diet that is up to two-thirds fat is acceptable and can, in fact, be critical to survival.

While consumption of small amounts of saturated fats is essential, initial meta-analyses (1997, 2003) found a high correlation between high consumption of such fats and coronary heart disease. Surprisingly, however, more recent meta-analyses (2009, 2010), based on cohort studies and on controlled, randomised trials, find a positive or neutral effect from shifting consumption from carbohydrate to saturated fats as a source of calories, and only a modest advantage for shifting from saturated to polyunsaturated fats (10% lower risk for 5% replacement).

Mayo Clinic has highlighted oils that are high in saturated fats, including coconut, palm oil and palm kernel oil. Those of lower amounts of saturated fats, and higher levels of unsaturated (preferably monounsaturated) fats like olive oil, peanut oil, canola oil, avocado, safflower, corn, sunflower, soy, mustard and cottonseed oils are generally healthier. The National Heart, Lung and Blood Institute and World Heart Federation have urged saturated fats be replaced with polyunsaturated and monounsaturated fats. The health body lists olive and canola oils as sources of monounsaturated oils while soybean and sunflower oils are rich with polyunsaturated fat. Results of research carried out in Costa Rica in 2005 suggest that consumption of non-hydrogenated unsaturated oils like soybean and sunflower are preferable to the consumption of palm oil.

Not all saturated fats have negative effects on cholesterol. Some studies indicate that Palmitic acid in palm oil does not behave like other saturated fats, and is neutral on cholesterol levels because it is equally distributed among the three “arms” of the triglyceride molecule. Further, it has been reported that palm oil consumption reduces blood cholesterol in comparison with other traditional sources of saturated fats such as coconut oil, dairy and animal fats.

Saturated fat is required by the body and brain to function properly. In fact, one study in Brazil compared the effects of soybean oil to coconut oil (a highly saturated fat) and found that while both groups showed a drop in BMI, the soybean oil group showed an increase in overall cholesterol (including a drop in HDL, the good cholesterol). The coconut oil group actually showed an increase in the HDL:LDL ratio (meaning there was more of the good cholesterol), as well as smaller waist sizes (something that was not shown in the soybean oil group. In 2007, scientists Kenneth C. Hayes and Pramod Khosla of Brandeis University and Wayne State University indicated that the focus of current research has shifted from saturated fats to individual fats and percentage of fatty acids (saturates, monounsaturates, polyunsaturates) in the diet. An adequate intake of both polyunsaturated and saturated fats is needed for the ideal LDL/HDL ratio in blood, as both contribute to the regulatory balance in lipoprotein metabolism.

Oils high in unsaturated fats may help to lower “bad” LDL cholesterol and may also raise “good” HDL cholesterol, though these effects are still under study. Peanut, cashew, and other nut-based oils may also present a hazard to persons with a nut allergy. A severe allergic reaction may cause anaphylactic shock and result in death.

Trans Fats

Unlike other dietary fats, trans fats are not essential, and they do not promote good health. The consumption of trans fats increases one’s risk of coronary heart disease by raising levels of “bad” LDL cholesterol and lowering levels of “good” HDL cholesterol. Trans fats from partially hydrogenated oils are more harmful than naturally occurring oils.

Several large studies indicate a link between consumption of high amounts of trans fat and coronary heart disease and possibly some other diseases. The United States Food and Drug Administration (FDA), the National Heart, Lung and Blood Institute and the American Heart Association (AHA) all have recommended limiting the intake of trans fats.

Cooking with Oil

Heating an oil changes its characteristics. Oils that are healthy at room temperature can become unhealthy when heated above certain temperatures. When choosing a cooking oil, it is important to match the oil's heat tolerance with the cooking method.

A 2001 parallel review of 20-year dietary fat studies in the United Kingdom, the United States of America, and Spain found that polyunsaturated oils like soya, canola, sunflower, and corn oil degrade easily to toxic compounds when heated. Prolonged consumption of burnt oils led to atherosclerosis, inflammatory joint disease, and development of birth defects. The scientists also questioned global health authorities' recommendation that large amounts of polyunsaturated fats be incorporated into the human diet without accompanying measures to ensure the protection of these fatty acids against heat- and oxidative-degradation.

Palm oil contains more saturated fats than canola oil, corn oil, linseed oil, soybean oil, safflower oil, and sunflower oil. Therefore, palm oil can withstand the high heat of deep frying and is resistant to oxidation compared to highly unsaturated vegetable oils. Since about 1900, palm oil has been increasingly incorporated into food by the global commercial food industry because it remains stable in deep frying or in baking at very high temperatures and for its high levels of natural antioxidants.

Oils that are suitable for high-temperature frying (above 230 °C/ 446 °F) because of their high smoke point:

- Avocado oil
- Corn oil
- Mustard oil
- Palm oil
- Peanut oil (marketed as "groundnut oil" in the UK and India)
- Rice bran oil
- Safflower oil
- Sesame oil (semi-refined)
- Soybean oil
- Sunflower oil.

Oils suitable for medium-temperature frying (above 190 °C/374 °F) include:

- Almond oil
- Cottonseed oil
- Diacylglycerol (DAG) oil
- Ghee, Clarified butter
- Grape seed oil
- Lard
- Olive oil (Virgin, and refined)
- Rapeseed oil, (marketed Canola oil in North America or, sometimes, simply "vegetable oil" in the UK)
- Mustard oil
- Walnut oil.

Unrefined oils should not be used for frying, but are safe for simmering.

Storing and Keeping Oil

Whether refined or not, all oils are sensitive to heat, light, and exposure to oxygen. Rancid oil has an unpleasant aroma and acrid taste, and its nutrient value is greatly diminished. To delay the development of rancid oil, a blanket of an inert gas, usually nitrogen, is applied to the vapour space in the storage container immediately after production. This is referred to as tank blanketing. Vitamin E oil is a natural antioxidant that can also be added to cooking oils to prevent rancidification.

All oils should be kept in a cool, dry place. Oils may thicken, but they will soon return to liquid if they stand at room temperature. To prevent negative effects of heat and light, oils should be removed from cold storage just long enough for use. Refined oils high in monounsaturated fats keep up to a year (olive oil will keep up to a few years), while those high in polyunsaturated fats keep about six months. Extra-virgin and virgin olive oils keep at least 9 months after opening. Other monounsaturated oils keep well up to eight months, while unrefined polyunsaturated oils will keep only about half as long. In contrast, saturated oils, such as coconut oil and palm oil, have much longer shelf lives and can be safely stored at room temperature. Their lack of polyunsaturated content causes them to be more stable.

Bibliography

Andrews J.: *The Domesticated Capsicums*, University of Texas Press. 1995.

Bodeker Gerard : *Medicinal Plants for Forest Conservation and Health Care*, Daya, Delhi, 2005.

Chaudhuri, A.B. : *Biodiversity Endangered: India's Threatened Wildlife and Medicinal Plants.* Scientific Publishers, Jodhpur, 2002.

Chevallier, Andrew : *Natural Health Encyclopedia of Herbal Medicine,* New York, DK Pub. Inc., 2000.

Crucefix D.: *Avocado Variety Selection for Export Development,* Roseau, CARDI, 1996.

Degras L.: *Yam: a Tropical Root Crop*, Wageningen, CTA/MacMillan, 1993.

Dudley, E.: *The Critical Villager: Beyond Community Participation*, London, Routledge, 1993.

Dutt, Ashwini : *An Introduction to Medicinal Plants*, Adhyayan Pub, Delhi, 2008.

Featherly H. I.: *Taxonomic Terminology of the Higher Plants*, USA, Iowa State College Press, 1954.

Fernie W T : *Medicinal Plants and Plant-Based Medicines*, Shree Pub, Delhi, 2008.

Giridhar A Kinhal : *Adaptive Management of Medicinal Plants and Non Timber Forest Products,*, BSMPS, Delhi, 2008.

Gopi K.S. : *Encyclopedia of Medicinal Plants Used in Homoeopathy*, Aiy Publications, Delhi, 2000.

Hartley W.: *A Checklist of Economic Plants in Australia*, Melbourne, C.S.I.R.O., 1979.

Hill, Clare: *The Ancient and Healing Art of Aromatherapy.* Ulysses Press, Berkeley, CA; 1997.

Janardhan K Reddy : *Advances in Medicinal Plants*, Universities Press, Delhi, 2007.

Janick, Jules: *Horticultural Science*, New York: W. H. Freeman and Company, 1986.

Julia F.: *Fruits of Warm Climates*, Miami, Julia F. Morton Publisher, 1987.

Kala C P : *Medicinal Plants of Indian Trans-Himalaya : Focus on Tibetan Use of Medicinal Resources*, BSMPS, Delhi, 2003.

Lawless, Julia: *Essential Oils: An Illustrated Guide.* Harpercollins, London, 2001.

Lininger, S.W. : *The Natural Pharmacy*, Prima Health, Rocklin, California, 1998 .

Mammen G.: *Studies of the Quantitative and Qualitative Attributes of Ginger Cultivars*, Calicut, Kasaragod, 1990.

McGilvery, Carole and Reed, Jimi: *The Complete Guide to Aromatherapy.* Lorenz Books, London, 2001.

Mills S : *The Essential Book of Herbal Medicine*, London: Arkana, 1991.

Nayar, M.P.: *Hot Spots of Endemic Plants of India, Nepal and Burma.* The Director, TBGRI, Trivandrum, 1996. 252 pp.

Oka H. I.: *Origin of Cultivated Rice*, Elsevier, Japan Scientific Societies Press, 1988.

Pareek O. P.: *Advances in Horticulture: Fruit Crops,* New Delhi, Malhotra Publishing House, 1993.

Percival John: *The Wheat Plant - A Monograph*, London, Duckworth & Co., 1921.

Polunin, O. and Stainton, A. : *Flowers of the Himalaya*, Oxford University Press, Delhi, India 1984.

Price, Shirley: *Aromatherapy for Common Ailments.* Simon & Schuster, New York, NY; 1991.

Pullaiah T. *: Encyclopaedia of World Medicinal Plants,* Regency, Delhi 2006.

Robbins, C. : *Herbalism: An Introductory Guide*, Parrallel, Bristol, 1995 .

Rutherford Lyn.: *A Gourmet's Book of Mushrooms & Truffles*, Sydney, Golden Press Pvt. Ltd., 1991.

Sharma Ravindra *: Medicinal Plants of India : An Encyclopaedia*, Daya Publishing, Delhi, 2003.

Shourie Arun *: Harvesting Our Souls*, Rupa, Delhi, 2001.

Thomas. G. S. : *Ornamental Shrubs, Climbers and Bamboos*, Murray 1992.

Trivedi, P.C. : *Indigenous Ethnomedicinal Plants*, Pointer Pub, Delhi, 2009.

Weiss, R.F. : *Herbal Medicine*, Beaconsfield Arcanum, Beaconsfield, 1991 .

Whistler, W.A. : *Polynesian Herbal Medicine*, National Tropical Garden, Kauai, Hawaii, 1992 .

Index

❑❑❑